This handy reference handbook describes the fundamental principles and procedures underlying the successful isolation of viable, functionally intact cells from mammalian endocrine tissues, and their maintenance as primary cultures. The cell types selected for coverage illustrate the diversity of endocrine tissues from which cells have been isolated, and the range of procedures which have been devised to ensure the optimal survival and behaviour of each cell type under study. Particular emphasis has been placed on the provision of detailed protocols describing, step by step, the manipulators necessary to establish differentiated and responsive cultures. The chapters have been prepared by authors having direct practical experience of the cell type concerned, and the reader is therefore provided with first-hand accounts on the background to each procedure, the avoidance of potential problems and pitfalls, and the fundamental question in endocrinology which may be addressed using each cell culture model.

Stephen Bidey is a lecturer in cell biology in the Endocrine Sciences Research Group of the University of Manchester. His principal research interest is the cell biology of the thyroid gland, especially with regard to the role of growth factors in goitrogenesis.

Handbooks in Practical Animal Cell Biology

Series editor:

Dr Ann Harris

Institute of Molecular Medicine, University of Oxford

The ability to grow cells in culture is an important and recognised part of biomedical research, but getting the culturing conditions correct for any particular type of cell is not always easy. These books aim to overcome this problem. The conditions necessary to culture different types of cell are clearly and simply explained in seven different volumes. Each volume covers a particular type of cell, and contains chapters by recognised experts explaining how to culture different lineages of the cell type. There is also a volume on general techniques in cell culturing. These practical handbooks are clearly essential reading for anyone who uses cell culture in the course of their research.

Already published in the series:

Epithelial cell culture, edited by A. Harris

Endothelial cell culture, edited by R. Bicknell

General techniques of cell culture, by M. Harrison and I. Rae

Marrow stromal cell culture, edited by J.N. Beresford and M.E. Owen

Forthcoming in the series:

Haemopoietic and lymphoid cell culture,
edited by M. Dallman and J. Lamb

Endocrine cell culture

Edited by
Stephen Bidey

CAMBRIDGE
UNIVERSITY PRESS

CAMBRIDGE
UNIVERSITY PRESS

University Printing House, Cambridge CB2 8BS, United Kingdom

Cambridge University Press is part of the University of Cambridge.

It furthers the University's mission by disseminating knowledge in the pursuit of education, learning and research at the highest international levels of excellence.

www.cambridge.org
Information on this title: www.cambridge.org/9780521595636

© Cambridge University Press 1998

First published 1998

A catalogue record for this publication is available from the British Library

Library of Congress Cataloguing in Publication data

Endocrine cell culture / edited by Stephen Bidey.
    p.   cm. – (Handbooks in practical animal cell biology)
Includes bibliographical references and index.
ISBN 0 521 59399 9 (alk.)
1. Endocrine glands – Histology.   2. Cell culture.   I. Bidey,
Stephen Peter, 1949–   .   II. Series.
QM576.E53   1998
611´.01894–dc21   98–14244   CIP

ISBN 978-0-521-59399-1 Hardback
ISBN 978-0-521-59563-6 Paperback

....

Contents

List of contributors	*page* vii	
Preface to the series	ix	
Acknowledgements	x	
Introduction STEPHEN BIDEY	1	
1	Parathyroid cells RONAL R. MACGREGOR	6
2	Ovarian granulosa and theca cells DENIS A. MAGOFFIN	20
3	Anterior pituitary cells SEON H. SHIN AND JOHN V. MILLIGAN	38
4	Pancreatic ß-cells LEONARD BEST	62
5	Adrenocortical and adrenomedullary cells MATTHIAS M. WEBER, CHRISTIAN FOTTNER AND DIETER ENGELHARDT	74
6	Leydig cells FRANK CHUZEL, HERVÉ LEJEUNE AND JOSÉ M. SAEZ	94
7	Thyroid follicular cells MARGARET C. EGGO	109
8	Hypothalamic cells HILARY E. MURRAY, DUNCAN McKENZIE AND GLENDA E. GILLIES	134
Index	151	

Contributors

Leonard Best
Department of Medicine, Stopford Building, University of Manchester, Manchester, UK

Stephen Bidey
Endocrine Sciences Research Group, University of Manchester, Manchester, UK

Frank Chuzel
INSERM-INRA 418, Hôpital Debrousse, Lyon, France

Margaret C. Eggo
Department of Medicine, Queen Elizabeth Medical Centre, University of Birmingham, Birmingham, UK

Dieter Engelhardt
Laboratory of Endocrine Research, Medical Department II, Klinikum Grosshadern, Ludwig Maximilians Universität, München, Germany

Christian Fottner
Laboratory of Endocrine Research, Medical Department II, Klinikum Grosshadern, Ludwig Maximilians Universität, München, Germany

Glenda E. Gillies
Department of Pharmacology, Charing Cross and Westminster Medical School, London, UK

Hervé Lejeune
INSERM-INRA 418, Hôpital Debrousse, Lyon, France

Ronal R. MacGregor
Department of Anatomy and Cell Biology, University of Kansas School of Medicine, Kansas City, Kansas, USA

Duncan McKenzie
Department of Pharmacology, Charing Cross and Westminster Medical School, London, UK

Denis A. Magoffin
Obstetrics and Gynaecolgy, Cedars-Sinai Medical Centre, Los Angeles, California, USA

John V. Milligan
Department of Physiology, Queen's University, Kingston, Ontario, Canada

Hilary E. Murray
Department of Pharmacology, Charing Cross and Westminster Medical School, London, UK

José M. Saez
INSERM-INRA 418, Hôpital Debrousse, Lyon, France

Seon H. Shin
Department of Physiology, Queen's University, Kingston, Ontario, Canada

Matthias M. Weber
Laboratory of Endocrine Research, Medical Department II, Klinikum Grosshadern, Ludwig Maximilians Universität, München, Germany

Preface to the series

The series Handbooks in Practical Animal Cell Biology was born out of a wish to provide scientists turning to cell biology, to answer specific biological questions, the same scope as those turning to molecular biology as a tool. Look on any molecular cell biology laboratory's bookshelf and you will find one or more multivolume works that provide excellent recipe books for molecular techniques. Practical cell biology normally has a much lower profile, usually with a few more theoretical books on the cell types being studied in that laboratory.

The aim of this series, then, is to provide a multivolume, recipe-book-style approach to cell biology. Individuals may wish to acquire one or more volumes for personal use. Laboratories are likely to find the whole series a valuable addition to the 'in house' technique base.

There is no doubt that a competent molecular cell biologist will need 'green fingers' and patience to succeed in the culture of many primary cell types. However, with our increasing knowledge of the molecular explanation for many complex biological processes, the need to study differentiated cell lineages *in vitro* becomes ever more fundamental to many research programmes. Many of the more tedious elements in cell biology have become less onerous due to the commercial availability of most reagents. Further, the element of 'witchcraft' involved in success in culturing particular primary cells has diminished as more individuals are successful. The chapters in each volume of the series are written by experts in the culture of each cell type. The specific aim of the series is to share that technical expertise with others. We, the editors and authors, wish you every success in achieving this.

ANN HARRIS

Acknowledgements

I must firstly express my gratitude to all of those who have contributed to this volume. Thanks are also due to Ann Harris at the University of Oxford for the invitation to prepare this volume, and to Barnaby Willitts and the editorial team at the Cambridge University Press who brought it to fruition.

S.B.

Introduction

Stephen Bidey

Endocrine cells of diverse origin share the common feature of synthesising peptide or steroid hormones which, following release into the bloodstream, exert biochemical effects on a multitude of different target cells throughout the body. Such actions are crucial to the response of the organism to changes in both internal and external environment. Certain hormones, such as adrenocorticotrophic hormone (ACTH) and thyroid-stimulating hormone (TSH) act exclusively on one target tissue, whereas others such as insulin are active on many cell types, and may elicit responses that differ between individual target tissues.

Our understanding of the physiological and molecular processes underlying hormone synthesis and action derives in large part from the *in vitro* study of endocrine cells in culture, which have been isolated from a wide range of endocrine tissues from numerous species. The development of reliable and stable cell culture models, which reproduce the essential features of the *in vivo* tissue environment, and in which derived cells can be manipulated, has been fundamental to our understanding of critical events in the growth, differentiation and function of each endocrine tissue. Significant insights have been gained, for example, into the mechanisms of hormone–receptor interaction, stimulus–response coupling, transcriptional regulation of prohormone synthesis, post-translational processing of prohormones, and the control of packaging and release of mature polypeptides. In many cases, the development of specialised culture systems has provided further information on the specific physical and chemical requirements of each cell lineage, enabling a further refinement in technique.

Primary cultures or cell lines?

As is the case with numerous other cell lineages, the culture of endocrine cells embraces both primary cultures and stable cell lines, the latter being

1

derived by viral transformation or selective cloning of primary cultures. Although many endocrine cell lines have been thus derived, it is notable that the growth, function and differentiation characteristics of these often demonstrate a marked divergence from the analogous primary cultures, as well as from the originating tissue. It is often observed, for example, that the terminal differentiation and function of endocrine cell lines are compromised, while cell proliferation, by definition, is maintained and frequently enhanced. A good example of this situation is seen in the case of thyroid follicular cells. Although many of the available thyroid cell lines retain the ability to process inorganic iodide and synthesise thyroglobulin in response to thyrotrophin (TSH), none actually secretes thyroid hormones. In contrast, in primary cultures of thyroid follicular cells maintained as three-dimensional follicular structures, synthesis and secretion of thyroid hormones are retained, while cell proliferation is minimal. For many studies of endocrine cell growth control, differentiation and function therefore, primary cultures continue to serve as the experimental system of choice.

For many cells of endocrine origin (e.g. thyroid follicular cells and adrenocortical cells, both of which are considered in this volume), isolation from freshly excised tissue is straightforward. The procedures and requirements for maintaining the derived cultures are well established, and such preparations retain their genetically determined patterns of differentiation and function for a finite period which may extend from days to a few weeks. Under suitably controlled conditions, reliable analyses of cell proliferation, differentiation and function can be undertaken with such models, and it is possible to investigate aspects of cell physiology and biochemistry that would be impossible to undertake in the intact animal. Several endocrine tissues, including the adrenal, pituitary and testis are composed of discrete and well-organised subpopulations of cells which are subject to different control mechanisms and synthesise hormones with differing target cell specificity. Purification of the individual cell types from such tissues, prior to the culture of each of these under controlled conditions, has considerably advanced our understanding of the behaviour and specific requirements of each cell type, that would be difficult to address in mixed or impure preparations. Conversely, studies of the paracrine relationships between differing cell types in a given endocrine tissue have been facilitated by the development of more complex culture models in which several cell populations may be manipulated independently.

In contrast to cells of adrenal and thyroid origin, for which robust and reliable culture procedures have been well established, the isolation of viable and responsive cells from a number of other endocrine tissues has proved problematical. In the case of the parathyroid glands, for example, the optimal requirements for survival and function of isolated cells in culture have only recently been appreciated, and appropriate models developed.

Endocrine disease

While cell culture models have afforded major advances in our under-standing of the control of growth, differentiation and function of cells from a wide range of normal endocrine tissues, the study of cells from patholog-ical endocrine tissues and endocrine tumours has yielded vital clues to the cellular and molecular basis of a wide range of endocrine pathologies. Moreover, in several cases, primary cell culture techniques have been successfully integrated with sophisticated and highly sensitive analytical and manipulative procedures in cellular and molecular physiology and biology, enabling information to be obtained from vanishingly small populations of cells.

Aims and scope of the volume

The primary objective of this volume is to consider the fundamental prin-ciples and problems associated with the isolation of viable, functionally intact cells from whole mammalian endocrine tissues, together with the techniques associated with their subsequent maintenance for experimental manipulation and study. We have not attempted to provide protocols for every type of endocrine cell, but in selecting those cell types for coverage, emphasis has been directed towards illustrating the diversity of endocrine tissues from which cells have been isolated and cultured, and the variety of techniques that have been used in the development of these *in vitro* models. The cell types selected for inclusion illustrate a broad spectrum of isolation and culture procedures, which demonstrate clear differences in the cell-specific requirements for tissue-dispersing enzymes, and variation in optimal dis-aggregation protocols, culture media, serum and growth supplements. For some of the endocrine cells considered, optimal survival and function have been found to be dependent upon highly specific conditions (e.g. mainte-nance on a supporting matrix, or as cellular aggregates), while others may be successfully maintained as conventional adherent monolayers.

In line with the philosophy and strategy of the other volumes in this series, detailed experimental protocols form a central component of each chapter. For each of the endocrine cell types considered, the selection of authors has been made in the knowledge that these individuals have direct practical experience of the cell type concerned, and are thus able to furnish the reader with first-hand information on the procedures and techniques involved, as well as potential problems and pitfalls which may be unique to a specific cell type.

The first chapter, on parathyroid cells, illustrates the problems of maintaining this cell type in a differentiated and responsive state in culture, and describes a novel system by which parathyroid cells may be maintained as three-dimensional aggregates, with stable and reproducible responses to Ca^{2+}, over a period of several weeks. Chapter 2, covering ovarian granulosa and theca cells, describes a culture model that has proved invaluable in studies of the regulation of steroidogenesis and ovarian follicle development, and has also enhanced our understanding of hormone signalling and gene transcription events. In Chapter 3, techniques are described for the isolation and study of cells from the anterior pituitary gland. Primary cultures of such cells have been used to demonstrate and dissect the neuroendocrine control of pituitary function by the hypothalamus, and pituitary feedback control by the secretions of other endocrine tissues. The chapter also includes protocols for assessment of cell function on the basis of hormone release and intracellular Ca^{2+} mobilisation. The isolation and culture of β-cells from the endocrine pancreas, as described in Chapter 4, have improved our understanding of the mechanisms of insulin secretion, and such model systems have provided valuable information on the interactive roles of ion transport and electrical events in glucose-induced insulin release. Chapter 5 deals with the individual cell types from the adrenal gland, and describes subtle differences in the procedures used to establish cultures from the cortical and medullary regions of this tissue. The former provides a good model for investigating the regulation of steroid hormone synthesis, metabolism and secretion, whereas the latter, as modified postganglionic sympathetic neurones, has proved invaluable in electrical stimulus–response studies. The chapter also describes subtle differences in the preparation and properties of cultures derived from tissue of bovine, rat, and human origin. Chapter 6 covers the isolation and culture of Leydig cells from porcine and human testes where, once again, differences are revealed in the optimal procedures for isolating analogous cells from differing species. Thyroid follicular cells are the subject of Chapter 7, where recognition of the importance of the spatial organisation of follicular cells within the thyroid epithelium has led to the development of models in which such features, along with a high level of functional integrity, are successfully reproduced. The final chapter deals with hypothalamic cells, the neuronal and glial cell subpopulations of which demonstrate a remarkable variation in structure and function. This chapter includes a section on the preparation, maintenance and study of tissue slices from defined hypothalamic regions, which have been used to complement primary cultures in investigations of cell–cell interactions.

Although some of the procedures described are unique to the specific cell lineage under consideration, others are applicable to a wider range of endocrine cells. In any event, the protocols described illustrate the diversity of both well-established and newly emerging techniques and experimental applications, as well as the fundamental principles and strategies upon which each procedure is based. It is hoped that, whatever type of endocrine cell the reader may wish to isolate, maintain and investigate, the range of protocols and wealth of expertise presented in this volume will facilitate the choice of an appropriate strategy. The procedures described may also suggest possibilities for the refinement of existing techniques and the development of entirely new approaches. Throughout this volume, it has been assumed that the reader will already have a basic working knowledge of cell culture, and is proficient in aseptic technique. Those without such a background are encouraged to read the general cell culture volume in this series first.

1

Parathyroid cells

Ronal R. MacGregor

Introduction

The primary function of the parathyroid glands is to secrete an 84-amino acid hormone called parathyroid hormone or parathormone (PTH) which serves to maintain the calcium concentration in the circulation. Adrenergic agents and the concentration of circulating, ionised calcium ($[Ca^{2+}]$) regulate PTH secretion in periods of seconds and minutes, while hormones such as insulin and vitamin D, in addition to $[Ca^{2+}]$, regulate secretion over more extended periods of hours to weeks. The need to examine agents that could regulate short-term PTH secretion has led to the adaptation of collagenase-based tissue digestion techniques for the isolation of parathyroid cells. A successful protocol for the isolation of bovine parathyroid cells was first reported by Brown, Hurwitz and Aurbach (1976). This method, based upon the use of bacterial collagenase and DNase, is still used today by many researchers for the digestion of bovine, porcine, and human parathyroid glands. The digestion yields a mixture of single cells and small clumps of cells.

In order to examine the secretory regulation of parathyroid cells over more extended periods, cell culture systems were required, and monolayer culture was the first such system considered (MacGregor, Cohn & Hamilton 1983a; LeBoff, Rennke & Brown 1983). Two problems arose during the development of monolayer culture systems. The first problem was that the clumps of cells obtained from the tissue digestion did not plate well on cultureware, but tended to remain in suspension, and then die. This has been handled in our laboratory by partially purifying the crude collagenase using Sephadex gel filtration (MacGregor *et al.*, 1983b).

The second problem encountered during monolayer culture of parathyroid cells was a loss of secretory responsiveness to changes in $[Ca^{2+}]$ that occurred within 3–6 days (LeBoff *et al.*,1983). The reason for this problem has only recently been traced to the loss of a receptor for divalent cations from the plasma membrane of cultured cells (Brown *et al.*, 1995; Mithal *et al.*, 1995). The reasons for the loss of this receptor have not yet been elucidated, and so the usefulness of monolayer cultures of parathyroid cells remains limited. The intractability of this problem has led us to examine possible alternative culture systems.

Examination of the research literature suggests that cultured parathyroid explants maintain calcium responsiveness better than any other system (Dietel & Dorn-Quint, 1980). Bovine parathyroid glands are extremely heterogeneous, however, and contain high and variable levels of collagenous matrix and fat. Production of sufficiently homogeneous natural explants for use in experiments therefore, is not feasible. We decided to determine if we could create a more uniform, explant-like aggregate from dispersed cells, and after a period of development devised a procedure that, in 3 days, yielded tissue-like aggregates that we have termed 'organoids' (Ridgeway, Hamilton & MacGregor 1986). The use of organoids has provided the stability and reproducibility needed for studies of the long-term actions of calcium (MacGregor *et al.*, 1988), insulin (Hinton & MacGregor, 1991), and vitamin D (Ridgeway & MacGregor, 1988).

Sources of cells and methods of isolation

A common procedure may be used to prepare parathyroid cells from tissue of slaughterhouse origin, regardless of their subsequent use as cell suspensions, monolayer cultures, or organoid cultures. There are two differences from the original procedure described by Brown *et al.* (1976). Firstly, the glands are sterilised before they are trimmed so that contamination of cultures can be prevented. The second difference is that the purity of the collagenase is higher and in some cases purified accessory proteases are added to the collagenase. Two types of digestion enzyme mixture have been used in our laboratory. The first is prepared from type I collagenase, while the second uses chromatographically purified collagenase.

Chemicals, reagents and supplies
Sterile culture plates and tubes (Falcon, Beckton-Dickinson & Co., NJ, USA)
Pressure dialysis unit and Diaflo™ PM-30 membranes (Amicon, Beverly, MA, USA)
Stadie–Riggs tissue slicing unit (Thomas Scientific, Swedesboro, NJ, USA)
Cell Production Roller Drum (Bellco Biotechnology, Vineland, NJ, USA)
Fetal bovine serum (Hyclone Laboratories Inc., Logan, UT, USA)
PhenoCen™ (Central Chemical Co. Inc., Kansas City, KS, USA)
Waymouth's MB752/1, buffers, insulin (Sigma Chemical Co., St Louis, MO, USA)

Collagenases CLS-1 and CLSPA, elastase, and papain (Worthington
Biochemical Corporation, Lakewood, NJ, USA)
Antibiotics (Gibco–BRL, Gaithersburg, MD, USA)

Preparation of digestion cocktails

The most economical enzyme preparation is prepared from type I collagenase by pressure dialysis. 1–5 g of crude enzyme can be processed using a 250
ml unit from Amicon. The enzyme is dissolved overnight at 4 °C in 100–200
ml of a buffer containing 1–2 mM calcium such as Hank's Balanced Salt
Solution (HBSS), or 0.2 M sodium acetate, (pH 7) with 1 mM $CaCl_2$ added.
The enzyme is placed in the dialysis cell that contains a PM-30 dialysis membrane, and 3–4 l of the same buffer is put into the dialysis reservoir. Pressure
dialysis at a nitrogen gas pressure of about 10 lb/in^2 may then be performed
in a cold room for 48–72 h with rapid stirring. 2–3 l of buffer should pass
through the membrane. The volume of the buffer in the chamber should then
be reduced by applying nitrogen pressure directly on the cell. A convenient
concentration of enzyme is 10 000 U/ml. At this concentration, each gram
of crude collagenase with a specific activity of 200 U/mg will provide 20 ml
of concentrate, 2.5 ml of which will be required for each 50 ml of digestion
medium. Particulate material in the enzyme solution may prevent filter sterilisation at this concentration, but the enzyme can be filtered after dilution into
digestion medium to attain the desired final concentration. This enzyme will
digest bovine parathyroid tissue safely and efficiently, albeit somewhat slower
than crude collagenase. If the digestions are slow, i.e. require over 5 h, the
amount of collagenase may be increased, or elastase can be added to the digestion medium at a final concentration of 100–200 μg/ml. Elastase aids tissue
digestion, but has no harmful effect on the cells.

The use of dialysed collagenase with or without elastase results in the isolation of cells in the form of small clumps. These clumps can be placed in
culture directly, but about 3 days are required for them to form a monolayer.
They can also be used as a fresh suspension in all types of incubation protocols for up to 6 h, and they can be used to form organoids. The isolated
parathyroid cells respond to changes in $[Ca^{2+}]$ in all three formats. If single
cell suspensions are required in order to form monolayers rapidly, activated
papain may be added to the digestion during the last 2 h at a final concentration of 200 μg/ml. These cells can be used for monolayers or organoid
formation, but do not respond to calcium until 24 h or more after isolation.

Digestion of parathyroid tissue can also be accomplished using a mixture
of purified collagenase (we have tested only CLSPA from Worthington) at
500–600 U/ml, and elastase at 200 μg/ml. The use of purified collagenase

is based on the untested assumption that proteolytic or lipolytic damage to plasma membranes, plasma membrane-bound receptors or other proteins exposed to the digestion medium will be minimised by the use of mixtures of pure enzymes with defined specificities. Accordingly, purified enzymes are used in order to promote reproducible and strong responses to chemical agents or to changes in $[Ca^{2+}]$, and in studies involving the investigation of changes in cytosolic calcium concentrations in the parathyroid cells.

Digestion of bovine parathyroid glands

The procedure described below is designed to digest glands of slaughterhouse animals, i.e. adult animals. It will also digest calf and human parathyroids, generally in a shorter time.

Protocol

1 Collect the glands at a slaughterhouse and put into ice-cold phosphate buffered saline containing penicillin, streptomycin, and gentamycin. Bring to the laboratory at 0 °C, drain, then transfer to a solution of 0.5% PhenoCen™ in saline at room temperature (some laboratories use 70% ethanol). Stir vigorously for 1–2 min in a sterile hood, in order to destroy bacterial contaminants. Rinse the glands four to six times with ice-cold saline, then transfer to ice-cold, sterile HBSS and store while trimming. All glassware and instruments should be sterile beyond this point.

2 Trim the glands on autoclaved paper towels in the sterile hood. With small surgical scissors, remove the fat and connective tissue and discard. Be careful not to crush the glandular tissue. The glands are small, about 100–300 mg. They vary in colour from yellow to red, and in texture from very fatty and full of connective tissue to relatively clean endocrine tissue. Place the trimmed glands in a sterile beaker containing cold, sterile HBSS. Keep on ice.

3 Slice the glands using a Stadie–Riggs tissue slicer. These are constructed of Plexiglass™ and cannot be autoclaved, so just rinse before use with 70% ethanol. Place the unit on autoclaved paper towels so that slices can be recovered. After placing a suitable number of glands in the slicer, apply pressure to the top of the unit with one hand, and with the other, move the blade continuously back and forth, rather than opening the unit after each slice. The slices will squeeze out from the sides and can be collected after processing all of the glands in the slicer. This process requires some practice in order to complete in a timely fashion. If you are unfamiliar with the Stadie–Riggs slicer, be very careful not to cut your fingers with the razor-sharp blade.

4 Transfer the slices to a 100 or 150 ml beaker of cold, sterile HBSS. After all slices are transferred, mince them with scissors until they can fit into the bore of a 10 ml serological pipette. After mincing, fill the beaker with HBSS, keep at 0 °C, and allow the slices to settle to the bottom of the beaker for 5 min. Remove the supernatant and discard, including gland slices that float. Refill the beaker with HBSS and repeat the settling and supernatant removal 3–4 times.

5 Remove as much HBSS as possible, then transfer the slices to an Erlenmeyer flask containing three to five gland volumes of digestion medium. The medium consists of Waymouth's MB752/1 medium, additionally buffered with 25 mM 3-(N-morpholino)propane-sulphonic acid (MOPS), containing 500 U/ml of collagenase. Cover and incubate in a shaking water bath at a speed just sufficient to send waves across the surface. Do not disturb for 30 min.

6 Beginning at 30 min, remove the flask to the sterile hood, and pipette the slices vigorously with a 10 ml serological pipette about 50–60 times. Repeat every 15 min. If difficulty is encountered, begin with a wide-mouth pipette. Initial pipetting may require the use of a 50 ml syringe connected to the pipette in order to reduce the size of the minced pieces of gland. As the pieces become smaller, reduce the vigour of the pipetting. Part-way through the digestion, the medium may become very viscous. This is normal; increase the shaking speed temporarily to compensate. Further digestion will degrade the viscous material. DNase is not necessary for the digestion protocol.

7 The entire digestion may require 4–5 h to complete. *It is important to complete the digestion* until no partially degraded connective tissue remains. If the digestion is not complete, connective tissue will precipitate when the cells are centrifuged and contaminate the pellets with fibrous and viscous material that will trap many cells.

8 Transfer the completed digest to centrifuge tubes; cap and centrifuge the cells at 50–100 g for 10 min. Carefully remove the supernatants and resuspend the cells in HBSS at room temperature. Recentrifuge for 5 min, then repeat the suspension and centrifugation steps until red cells are no longer observed at the top of the pellets.

9 Suspend the cells in culture medium and remove representative samples to be counted. Because the cells are present as clumps, trypsin-EDTA digestion is necessary prior to cell counting. The yield of cells should be $2 \times 10^7 - 4 \times 10^7$ per gram of trimmed glands.

Since the digestion is fairly extended, and the glands may reach the laboratory at different times, there are steps at which the procedure may be stopped and the tissue or cells kept overnight. The cells may be kept overnight at 4 °C after slicing but prior to digestion, or part-way through the digestion. They may also be kept overnight at either 4 °C or at room temperature after they have been isolated, rinsed, and placed in culture medium. To keep all of the cells overnight at room temperature carries some risk of bacterial contamination. Cells that are kept in that manner, however, give excellent secretory responses, suggesting that damaged receptors may be replaced during the overnight storage. This is consistent with our finding that cells isolated in the presence of papain do not respond to calcium as fresh suspensions, but do so normally after 24 h or more of monolayer culture (R.R. MacGregor, unpublished results).

Establishment and maintenance of cultures

Monolayer culture of parathyroid cells

Parathyroid cells isolated as described above adhere well to standard Falcon or Primaria cultureware in the presence of 5% fetal bovine serum. Several culture media are satisfactory; we have always favoured those that permit culture in an air atmosphere such as Waymouth's MB752/1 (Fig. 1.1). Parathyroid cells require supplementation with insulin (10^{-8} M) in order to maintain a stable, basal rate of PTH secretion (MacGregor et al., 1983b; Hinton & MacGregor, 1991). Seeding efficiencies are essentially quantitative. Since parathyroid cells are small and terminally differentiated, we favour seeding at relatively high densities ($5 \times 10^5 - 7 \times 10^5$ cells/cm²). Although culture can be continued in the presence of serum, this is not required after overnight adhesion of the cells, and in some cases the morphological appearance of the cells in monolayer suggests that serum may be harmful. The cells respond well to changes in [Ca^{2+}] in serum-free media containing 0.1% bovine serum albumin, which is added to inhibit non-specific adhesion of secreted molecules to the plastic culture dishes. Because of the problem of functional stability of these cultures mentioned above, experiments should be timed for completion within 3–5 days.

Organoid culture of parathyroid cells

Parathyroid organoids are a convenient alternative to cultures in a monolayer format, and since calcium sensitivity is retained for up to 3 weeks, this allows sufficient time to test the effects of agents and to examine reversibility after their removal (Hinton & MacGregor, 1991). In addition, organoid formation is straightforward. Organoids can be formed in either small (12×75

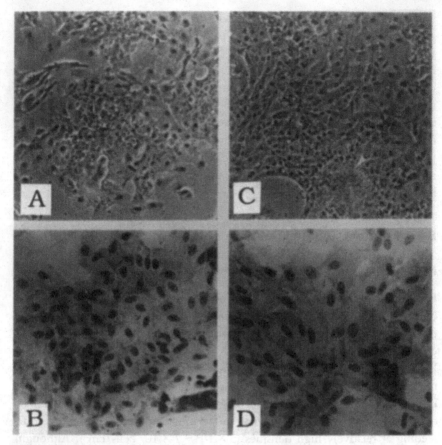

Fig. 1.1. Light microscopy of monolayer cultures of adult bovine parathyroid cells. Cells were cultured for 3 (A, B) or 7 (C, D) days in Waymouth's MB752/1 medium containing 5% fetal bovine serum supplemented with insulin, hydrocortisone, transferrin and epidermal growth factor. A, C, phase contrast (×175 magnification); B, D, bright field (×350 magnification). (From MacGregor, et al. 1983a, by permission of Elsevier Science Ireland Ltd.),

mm) or large (17 × 100 mm) sterile polystyrene tubes with caps, if the proper roller drum rack is available.

Protocol

1 Suspend freshly isolated parathyroid cells in serum-free culture medium such as MOPS-buffered Waymouth's MB752/1 medium at a concentration of 10^6 cells per ml. Into 12 × 75 mm tubes, transfer 0.75–1 ml of cells, and into 17 × 100 mm tubes, transfer 1.5–2 ml. Cap the tubes and then centrifuge for 5 min at approximately 2000 g at room temperature, or until firm pellets are formed.

2 Gently remove the tubes from the centrifuge, and without disturbing the circular pellets, transfer the tubes to the Bellco roller drum inside a 37 °C culture incubator. Rotate the roller at 2–3 rpm and do not disturb for 2 days. Experiments can be initiated after 2 or 3 days.

Medium should be changed every 2–3 days, or if medium begins to acidify. Since organoids continue to contract during culture, some care must be taken not to remove small organoids during medium removal. The centrifuged format is favoured because the organoids have a two-dimensional character, i.e. shorter diffusion distances, and uniform size and shape.

In our initial studies, (Ridgeway et al., 1986), organoids were formed by spontaneous aggregation in the roller drum. When formed in this manner, they tend to be cigar-shaped and less uniform than those formed using the centrifuge. In both cases the organoids shrink over the first 3 days until an explant-like structure is formed (Fig. 1.2). Ultrastructurally, the parathyroid cells within the organoids resemble glandular cells; the membranes are interdigitated, Golgi are visible, and prosecretory and secretory granules are scattered in the cytoplasm (Fig. 1.3).

Figure 1.4 illustrates the effect of culture time on calcium responsiveness in the steady state. In this experiment, spontaneously formed organoids were maintained at 1.0, 1.35 or 1.7 mM total [Ca] (calcium within the medium is about 85% ionised) for the entire experiment, and samples were removed for PTH radioimmunoassay 24 h after each medium change. In separate experiments (data not shown), it has been demonstrated that organoids are capable of responding to changes in [Ca^{2+}] within minutes (Ridgeway et al., 1986). Figure 1.5 shows that, after culture with 1.4 mM Ca for 6 days, organoids are able to modulate the rates of secretion of PTH in response to differences in [Ca^{2+}] to a degree that is similar to that exhibited by freshly isolated parathyroid cells.

Concluding remarks

Techniques have been described for preparing collagenase-based digestion cocktails that are superior to crude collagenases in their ability to disperse bovine parathyroid glands in such a way that cell yields are high and reproducible. The cells produced by these digestions respond to different [Ca^{2+}] strongly and reproducibly and plate efficiently into cultureware. In addition, modifications of digestion cocktails have been described that permit isolation of cells either as clumps or as single cells. A digestion cocktail that permits the use of chromatographically purified collagenase combined with elastase has been described that avoids completely the use of uncharacterised

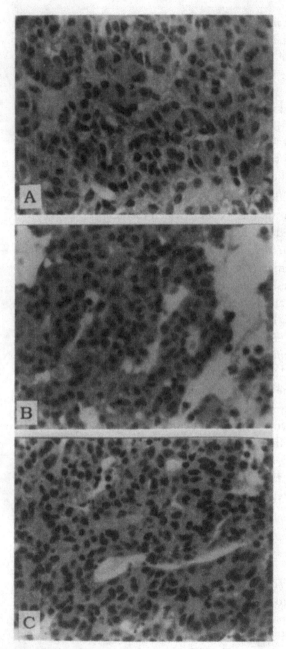

Fig. 1.2. Light micrographs of bovine parathyroid organoids and intact bovine parathyroid tissue (×130 magnification). A, Bovine parathyroid tissue; the semiacinar cellular architecture is apparent. B, Organoids of bovine parathyroid cells after 2 days in culture. Cells have formed multicellular aggregates and cord-like structures. C, Organoids after 6 days of culture. The cell mass has compacted and appears similar to the original tissue. (From Ridgeway et al., 1986, by permission of In vitro, Texas A&M University-IBT.)

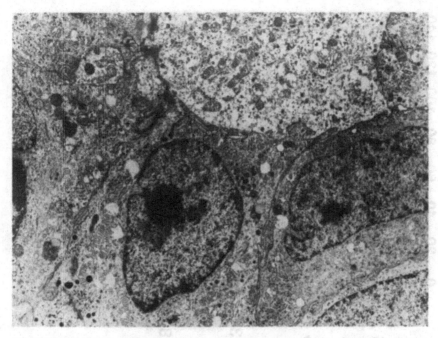

Fig. 1.3. Electron micrograph of a bovine parathyroid organoid after 9 days of culture (×5340 magnification). After 3 days of culture in Waymouth's MB752/1 medium (0.8 mM Ca^{2+}), [Ca^{2+}] was raised to 1.4 mM, a concentration close to physiological [Ca^{2+}]. (From Ridgeway et al., 1986, by permission of In vitro, Texas A&M University-IBT.)

collagenases. Finally, procedures for placing parathyroid cells in culture have been described that permit the formation of explant-like organoids, which can be used for studies extending for several weeks, and are amenable to morphological methods of experimental examination.

Troubleshooting

A number of potential difficulties and their solutions have already been described in the text, but descriptions of the most frequent difficulties are listed below:

1 *Minced parathyroid tissue cannot be pipetted.* Parathyroid tissue is very dense and will not deform to enter a pipette. It must be sliced thinly and minced with scissors to give pieces small enough to enter the pipette. This requires some time to accomplish. In some cases, pipetting can be initiated with wide-bore pipettes, but this extends digestion time. Initial pipetting is best performed by attaching a 10 ml serological pipette to a 50 ml syringe with

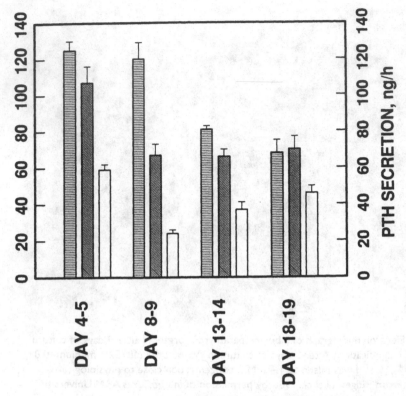

Fig. 1.4. The effects of culture time on the ability of calcium to modulate steady-state PTH secretion from bovine parathyroid organoids. Organoids were separately cultured in Waymouth's MB752/1 medium containing 10^{-6} M insulin and other additions as described (Ridgeway *et al.*, 1986). Total [Ca^{2+}] (Ca^{2+} was about 85% ionised in the culture medium) was 0.8 mM for day 0–2, and then was changed to 1.0 (horizontal hatch), 1.35 (diagonal hatch), or 1.7 mM (open bars). Samples representing 24 h of secretion into fresh media were removed on the days shown, and subjected to radioimmunoassay for PTH using antiserum 1811 (MacGregor, Cohn & Hamilton, 1983). Results are redrawn from a portion of Table 1 of Ridgeway *et al.*, 1986. The data for each [Ca^{2+}] are derived from successive samples of the same organoids. $n=6$ per point. All averages at 1.7 mM Ca are significantly different ($p<0.01$) from the corresponding data at 1.35 mM Ca. The same is true for averages at 1.35 vs. 1.0 mM Ca except for those on day 19.

a piece of vacuum tubing, and using the strong vacuum of the syringe to draw in the tissue. After the tissue pieces begin to enter the pipette easily, switch to a standard pipette bulb.

2 *The glands don't digest or digest too slowly.* This results from inadequate amounts of enzyme activity. Using the PM-30 dialysed enzyme, up to 600 U/ml (this assumes theoretical yields of activity after PM-30 filtration)

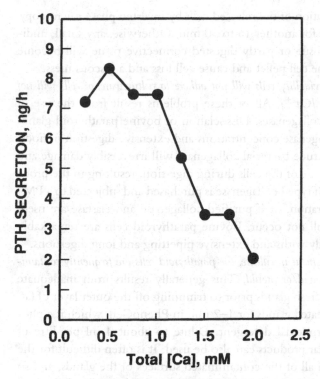

Fig. 1.5. Effects of [Ca^{2+}] on parathyroid hormone (PTH) secretion. Bovine parathyroid organoids were separately cultured for 6 days at a total [Ca^{2+}] of 1.4 mM. They were then incubated for 18 h in the same medium but with [Ca^{2+}] of 0.25, 0.5, 0.75, 1.0, 1.25, 1.5, 1.75, and 2.0 mM. Samples of the media were subjected to radioimmunoassay for PTH. Data are averages of five organoids per point, with the SEM for each point less than 10% of the average value.

may be necessary for digestion of glands from beef cattle (less for calf or human glands). In addition, if the original crude collagenase contains low levels of other necessary enzyme activities (an example would be collagenase 3 from Worthington Biochemicals), it may be necessary to add elastase to the digestion. Normally, collagenase I will work well after PM-30 filtration. The use of purified collagenase (e.g. CLSPA from Worthington) combined with elastase is normally only desirable if digestions need to be as gentle as possible, such as when cytosolic [Ca^{2+}] is to be measured. It is more expensive, and about 500 U/ml of digestion medium are required in order to complete digestion in 4–5 h.

3 *When the completed digest is first centrifuged to recover the cells, the pellets contain gummy, gelatinous material that traps many cells and reduces the yield.* This results from incomplete digestion. After the digestion appears to be complete, as

judged by visualisation of the released cells by inverted phase microscopy, continue to digest for another 15 to 30 min. Otherwise, any small, undigested pieces of tissue or partly digested connective tissue will become concentrated in the cell pellet and cause cell loss and a viscous mess.

4 *Low cell yield, low viability, cells will not adhere to cultureware, or cells will not respond to different [Ca^{2+}]*. All of these problems result from the use of untreated, crude collagenases. Dissociation of bovine parathyroid glands requires high collagenase concentrations and extensive digestion periods. Many batches of crude bacterial collagenase will irreversibly damage and kill a high percentage of the cells during digestion, resulting in the problems listed above. If type I collagenase is purchased and subjected to a PM-30 membrane filtration, or if purified collagenase and elastase are used, these problems will not occur. Bovine parathyroid cells are remarkably hardy and can easily withstand extensive pipetting and long digestions.

5 *Following their placement in culture, the parathyroid cells are frequently contaminated with bacteria and/or mould*. This generally results from inadequate sterilisation of the fresh glands prior to trimming off the outer layer of fat. We have always treated glands for 1–2 min in PhenoCen, which is a phenolic, hospital germicidal detergent. Dilute to about 4 ml per litre of saline. Other similar products can also be used. It is often difficult for the detergent to reach all of the contaminated surfaces of the glands, and so they must be stirred vigorously. In addition, the glands may be difficult to identify at the slaughterhouse, and so some collectors cut partially into them in order to see the glandular tissue. The glands can close over these cuts, trapping bacteria inside. It may be necessary to examine the fresh glands and to slice completely through any that have been cut prior to sterilisation.

References

Brown, A.J., Zhong M., Ritter, C., Brown, E.M. & Slatopolski, E. (1995). Loss of calcium responsiveness in cultured bovine parathyroid cells is associated with decreased calcium receptor expression. *Biochem. Biophys. Res. Commun.*, **212**, 861–7.

Brown, E.M., Hurwitz, S. & Aurbach G.D. (1976). Preparation of viable isolated bovine parathyroid cells. *Endocrinology*, **99**, 1582–8.

Dietel, M. & Dorn-Quint, G. (1980). By-pass secretion of human parathyroid adenomas. *Lab. Invest.*, **43**, 116–25.

Hinton, D.A. & MacGregor, R.R. (1991). Effects of insulin on the synthesis, intracellular degradation, and secretion of parathormone. *Endocrinology*, **128**, 488–95.

LeBoff, M.S., Rennke, H.G. & Brown, E.M., (1983). Abnormal regulation of parathyroid cell secretion and proliferation in primary cultures of bovine parathyroid cells. *Endocrinology*, **113**, 277–84.

MacGregor, R.R., Cohn, D.V. & Hamilton, J.W. (1983*a*). The content of carboxyl-terminal fragments of parathormone in extracts of fresh bovine parathyroids. *Endocrinology*, **112**, 1019–25.

MacGregor, R.R., Sarras, M.P. Jr., Houle, A. & Cohn, D.V. (1983*b*). Primary mono-layer cell culture of bovine parathyroids: effects of calcium, isoproterenol and growth factors. *Mol. Cell. Endocrinol.*, **30**, 313–28.

MacGregor, R.R., Hinton, D.A. & Ridgeway, R.D. (1988). Effects of calcium on synthesis and secretion of parathyroid hormone and secretory protein I. *Am. J. Physiol.*, **255** (*Endocrinology and Metabolism 18*), E299–305.

Mithal, A., Kifor, O., Kifor, I., Vassilev, P., Butters, R., Krapcho, K., Simin, R., Fuller, F., Hebert, S.C. & Brown, E.M. (1995). The reduced responsiveness of cultured bovine parathyroid cells to extracellular Ca^{2+} is associated with marked reduction in the expression of extracellular Ca^{2+}-sensing messenger ribonucleic acid and protein. *Endocrinology*, **136**, 3087–92.

Ridgeway, R.D. & MacGregor, R.R. (1988). Opposite effects of 1,25$(OH)_2D_3$ on synthesis and release of PTH compared with secretory protein I. *Am. J. Physiol.*, **254** (*Endocrinology and Metabolism 17*), E279–86.

Ridgeway, R.D., Hamilton, J.W. & MacGregor, R.R. (1986). Characteristics of bovine parathyroid cell organoids in culture. *In vitro Cell. Develop. Biol.* **22**, 91–9.

2

Ovarian granulosa and theca cells

Denis A. Magoffin

Introduction and applications

The discovery that the mammalian ovary is the principal site of estrogen production was a relatively straightforward finding. The ovary is a complex organ consisting of a large number of follicles in various stages of development as well as interfollicular stroma and a variable number of corpora lutea. Each of the ovarian follicles contains a single oocyte surrounded by multiple layers of cuboidal granulosa cells (GC). The Graafian follicles contain a fluid-filled antrum whereas the immature (preantral) follicles do not. The granulosa layers are enclosed within a basal lamina that physically isolates the interior of the follicle from the surrounding stromal tissues (Fig. 2.1). Adjacent to the basal lamina are several layers of endocrine cells comprising the theca interna (TIC) and then a few layers of contractile cells in the theca externa. It was not until techniques were developed to study isolated components of the ovary that it was discovered that estradiol production requires co-operation between the TIC and the GC.

The ovarian follicle undergoes a tightly regulated programme of growth and differentiation. Studies using a variety of methods, including *in vitro* cell cultures, have shown that the TIC differentiate in early preantral follicle development by expressing receptors for luteinising hormone (LH) and basal levels of the steroidogenic enzymes required for androgen production, namely cholesterol side chain cleavage cytochrome P450 ($P450_{SCC}$), 3β-hydroxysteroid dehydrogenase (3β-HSD) and 17α-hydroxylase/C_{17-20} lyase cytochrome P450 ($P450_{17\alpha}$). During the course of follicle development, LH plays a primary role in stimulating increasing amounts of the steroidogenic enzymes. The GC express receptors for follicle-stimulating hormone (FSH) early in development, but do not express steroidogenic enzymes until the early antral stage of follicle development. GC from mature antral follicles contain aromatase cytochrome P450 ($P450_{AROM}$) as well as $P450_{SCC}$ and 3β-

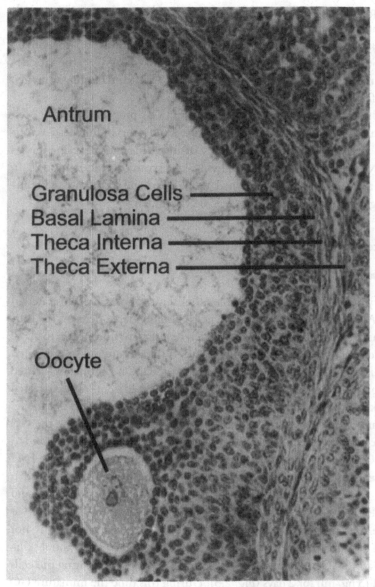

Fig. 2.1. Morphology of the ovarian Graafian follicle. The Graafian follicle contains a fluid-filled antrum and a single oocyte surrounded by cuboidal granulosa cells. The granulosa cells are enclosed within a basal lamina which is surrounded by several layers of theca interna and finally by the theca externa. There are convenient dissection planes at the basal lamina and between the theca interna and theca externa.

HSD. Expression of $P450_{AROM}$ is specific to the GC in most species including the rat and human, but other species such as the pig and horse express small amounts of $P450_{AROM}$ in the TIC as well. Since the TIC do not contain large amounts of $P450_{AROM}$ and the GC are devoid of $P450_{17\alpha}$, the production of large amounts of estradiol requires the TIC to make the androgen substrate for aromatisation by the GC.

Over the past several years, isolated TIC and GC preparations have been used to study the production of local regulators of ovarian follicle development and steroidogenesis. A wide variety of autocrine and paracrine mechanisms have been examined using individual cells and co-culture models. In addition, a great deal has been learned about hormone signalling and regulation of gene transcription using ovarian cell models.

Characterisation and morphology

Granulosa cells

In primary culture, GC grow with an epithelial morphology, as illustrated in Fig. 2.2 (A), but the appearance of the cells is highly dependent on the substrate. On uncoated plastic, the GC will secrete sufficient fibronectin to allow the cells to attach to the substrate and spread. If the culture dish is precoated with serum, the GC will readily attach and spread. On fibronectin-coated dishes the GC are extremely flattened and tightly adherent. When treated with forskolin (25 μM), the GC will round up. Characterisation of the GC will depend on the stage of development of the follicle from which the GC were collected and the hormonal treatments they may have been exposed to *in vivo*. GC from all stages of development express FSH receptors. The FSH receptor is expressed exclusively in GC within the ovarian follicle and is therefore a good specific marker. In many species such as rodents and humans, $P450_{AROM}$ is a specific marker of the GC. In other species such as the horse and pig, the TIC express some $P450_{AROM}$. $P450_{AROM}$ is not expressed in immature GC but is abundantly expressed in large antral follicles. Immature GC can be induced to express $P450_{AROM}$ by treating the cells with FSH (3 ng/ml) for 2 days (Fig. 2.3). During this time the immature GC also express $P450_{SCC}$, 3β-HSD and LH receptors, although none of these is specific for GC.

Theca cells

Thecal interstitial cells exhibit a fibroblastic morphology in culture (Fig. 2.2 (B)) regardless of the substrate on which they are cultured. They may tend

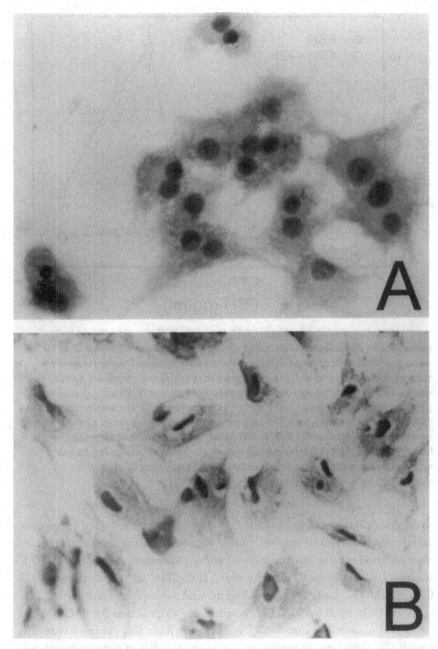

Fig. 2.2. Human granulosa and theca cells in culture. Isolated granulosa cells (A) or theca cells (B) were plated in serum-free McCoy's 5A medium containing 100 U/ml penicillin, 100 μg/ml streptomycin sulphate, and 2 mM L-glutamine. After 48 h of culture at 37°C in 95% air, 5% CO_2, the cells were fixed with neutral buffered formalin, stained with haematoxylin and eosin, and photographed (×125 magnification).

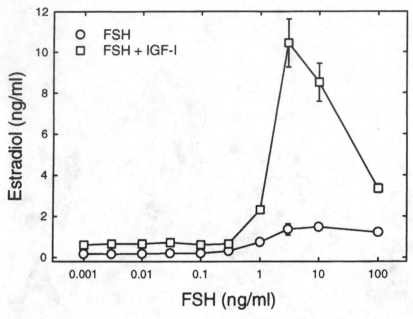

Fig. 2.3. Follicle-stimulating hormone (FSH) stimulation of granulosa cell estradiol production in the presence and absence of insulin-like growth factor-I (IGF-I). 60 000 viable granulosa cells obtained from immature intact Sprague-Dawley rats were cultured for 2 days in 96-well tissue culture plates (each well containing a total volume of 200 μl of McCoy's 5a medium supplemented with 100 U/ml penicillin, 100 μg/ml streptomycin sulphate and 2 mM L-glutamine) at 37°C in a water-saturated 95% air / 5% CO_2 atmosphere. The cells were cultured in the presence of increasing concentrations of ovine FSH (0.001–100 ng/ml) with and without recombinant human IGF-I (30 ng/ml). After 2 days of culture the medium was collected and estradiol in the medium was measured by radioimmunoassay. (Data are the mean±SEM.)

to round up in response to forskolin, but there is a variable degree of response which precludes this as a good method for determining purity of the cultures. A good specific marker for TIC is the expression of $P450_{17\alpha}$. $P450_{17\alpha}$ is expressed only in the TIC of the ovary. If the TIC are obtained from small follicles, the GC are not yet steroidogenic and a convenient marker for TIC is 3β–HSD. In small follicles, LH receptors are exclusive to the TIC and can be a good marker, but mature GC and corpora lutea also express LH receptors. If GC from large or medium-sized antral follicles or corpora luteal cells are potential contaminants, LH receptors will not be a specific marker. In immature rats, LH receptors are an excellent marker for TIC. A characteristic functional response of the TIC is the production of androgens in response to LH stimulation. A good TIC culture can be stimulated by LH

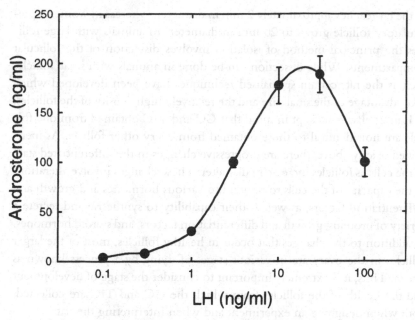

Fig. 2.4. Luteinising hormone (LH) stimulation of rat theca-interstitial cell androgen production. 20 000 viable theca-interstitial cells purified by Percoll density gradient centrifugation were cultured in 96-well tissue culture plates (each well containing a total volume of 200 μl of McCoy's 5a medium supplemented with 100 U/ml penicillin, 100 μg/ml streptomycin sulphate and 2 mM L-glutamine) at 37°C in a water-saturated atmosphere of 95% air / 5% CO_2. The cells were cultured in the presence of increasing concentrations of ovine LH (0.01–100 ng/ml). After 2 days of culture the medium was collected and androsterone in the medium was measured by radioimmunoassay. (Data are the mean±SEM.)

to increase androgen production many fold (Fig. 2.4). In adult mammals, the principal androgen produced by the TIC is androstenedione whereas in immature rats the principal androgen is androsterone. During prepubertal development high levels of 5α-reductase and 3α-reductase are expressed that metabolise androstenedione to androsterone. At puberty, these enzymes decline in the rat and androstenedione is the major product.

Sources of cells

GC and TIC can be cultured from a variety of species including the mouse, rat, hamster, rabbit, pig, sheep, cow and horse, in addition to human females. The methods for isolating GC and TIC are similar in most species, any differences being related to the variation in sizes of the ovaries and in the diameter of follicles in different species. For example, a preovulatory follicle in the rat reaches approximately 2 mm in diameter, whereas in women a pre-ovulatory follicle grows to 20 mm in diameter. In animals with large follicles the principal method of isolation involves dissection of the follicular compartments. While dissection can be done in animals with small follicles such as the rat, certain specialised techniques have been developed which take advantage of the small size and the relatively high density of the follicles.

It must always be kept in mind that GC and TIC obtained from one follicle are not identical to those obtained from every other follicle. As mentioned briefly above, there are progressive changes in the differentiated state of the cells as follicles increase in diameter. These changes involve alterations in the capacity of the cells to respond to various hormones and growth and differentiation factors, as well as their capability to synthesise and secrete a variety of proteins, growth and differentiation factors, and steroid hormones. In addition to the changes that occur in healthy follicles, most of the larger follicles in the ovary are in various stages of dying by a process known as atresia. Thus, it is extremely important to consider the stage of development and the health of the follicles from which the GC and TIC are collected, both when designing an experiment and when interpreting the data.

Methods of isolation

Granulosa cells

There are three principal methods of isolating GC from ovarian follicles; microdissection, aspiration, and follicle puncture. Microdissection can be used on follicles of any species and of all sizes. The only practical limitations are (i) that microdissection becomes increasingly difficult as the size of the follicle decreases and (ii) that the ovary must be removed from the animal. A common technique used with women from whom ovaries cannot be removed is to aspirate the contents of the follicle and isolate the GC from the aspirate. This is particularly common with hyperstimulated ovaries from which the oocytes are being retrieved for use in assisted reproductive technologies. Follicle puncture is the method of choice for species with very small follicles such as the rat, but will not work effectively with larger follicles.

Materials
H-199 (HEPES buffered Medium-199): Medium-199 with Hank's salts, L-glutamine, with 25 mM HEPES buffer (GIBCO, cat. no. 12350-039)
H-199B: Dissolve 100 mg of bovine serum albumin in 100 ml H-199. Prepare fresh and sterilise by membrane filtration (0.2 μM).
Percoll (Pharmacia, cat. no. 17-0891-01)

Microdissection

Protocol

1 Remove the ovary as quickly as possible after sacrificing the animal. Take care to remove the ovary using aseptic technique. Place the tissue immediately into ice-cold H-199B. The dissection is performed in HEPES buffered medium because the HEPES buffer will maintain a reasonably physiological pH in an air atmosphere whereas the pH of bicarbonate-buffered tissue culture media will rapidly rise.

2 Thoroughly wash off all blood from the surface of the ovary. This is important because the granulosa cells will be collected from the medium and residual blood cells will contaminate the GC.

3 Place the washed ovary into the lid of a sterile petri dish containing a small amount (~10 ml) of H-199B.

4 Insert a syringe with a 25-gauge needle into a follicle and gently aspirate all of the follicular fluid. If the follicles are less than 11 mm in diameter, a 1 ml tuberculin syringe can be used, however larger follicles will require a 5 ml syringe. Using the follicular fluid volume to estimate the follicular diameter is more reliable than trying to physically measure the follicle since follicles are frequently not perfectly spherical, but one must obtain all of the follicular fluid from the follicle.

5 Centrifuge the follicular fluid 5 min at 250 g to collect the GC and save the pellet on ice. Freeze the follicular fluid, if desired, until analysis for steroid hormones or other components of interest is performed. If the volume is large, it is advisable to aliquot the fluid from each follicle to minimise freeze–thaw cycles.

6 Using a scalpel, make a small incision in the follicle wall large enough to insert an inoculating loop into the follicle. It is wise to collect the tissue and fluid samples from each follicle at the same time rather than aspirating the fluid from all of the follicles prior to dissection. Once the follicle has collapsed, it is more difficult to locate later, especially if it is small.

7 Gently scrape the GC from the follicle wall with the inoculating loop. Vigorous scraping will kill large numbers of GC.

8 Gently flush the GC from the follicle with a small amount of medium. Some of the GC will spill into the medium in the petri dish. Collect them with a sterile Pasteur pipette and pool the GC with the pellet obtained from the follicular fluid.

9 Wash the cells with tissue culture medium and resuspend the pellet in 1 ml of tissue culture medium.

10 Count the cells in a haemocytometer using trypan blue exclusion to determine viability. The procedure will have gone well if >50% of the

GC are viable. Should there be significant contamination with red blood cells, purify the GC through 50% Percoll as described for the aspiration protocol. Spinning 0.5 ml of GC over 1.0 ml of 50% Percoll in a 1.7 ml microcentrifuge tube works well when there are small numbers of GC.

11 If it is desired to collect TIC, proceed to the microdissection protocol for isolating TIC.

Follicle aspiration of GC

Protocol

1 Locate the antral follicle on the surface of the ovary. Insert a 25 gauge needle into an avascular region of the follicle and aspirate the follicular fluid. Flush as many GC from the follicle as possible with sterile balanced salt solution. Gently scraping the follicle wall can help increase the yield of cells. Collect the GC into a heparinised sterile tube on ice. The follicle aspirates will generally have significant blood contamination. This does not present a problem unless coagulation occurs. If clots form, the GC will be bound together with red blood cells into clumps that prevent purification of the GC.

2 Centrifuge the aspirates at 250 g for 10 min. Resuspend the GC in a suitable volume of H-199B. The volume of medium used for resuspension will vary depending on the level of blood contamination and the volume of the original aspirates. When purifying GC on Percoll gradients, the GC will accumulate at the interface of the 50% Percoll layer. If there are too many cells at the interface, they will trap red blood cells at the interface. If too much medium is used the yield will be poor. Start with a volume approximately one-tenth of the original aspirated volume.

3 Prepare 50% Percoll by mixing an equal volume of Percoll and H-199B in a sterile tube. The Percoll gradients are prepared by first pipetting 50% Percoll into sterile centrifuge tubes and then carefully layering an equal volume of cell suspension onto the top of the 50% Percoll layer taking care to avoid mixing of the two layers. Use as many centrifuge tubes as necessary to purify the volume required. It is easier to prevent mixing by using tubes with smaller diameters.

4 Centrifuge the tubes at 450 g for 20 min at 4 °C. Turn off the brake on the centrifuge and let the rotor coast to a stop to prevent disturbing the gradient after centrifugation is complete. After centrifugation the granulosa luteal cells should have formed a band at the interface of the 50% Percoll layer and the red blood cells should pellet at the bottom of the tube.

5 Using a sterile Pasteur pipette, collect the GC by aspiration into a 15 ml sterile centrifuge tube.

6 Add an equal volume of tissue culture medium, mix gently and centrifuge at 250 g for 5 min at 4 °C.

7 Gently resuspend the pellet in a known volume of tissue culture medium, which will yield a concentration of approximately $10 \times 10^6 - 30 \times 10^6$ cells/ml. Count the cells in a haemocytometer using trypan blue exclusion to determine viability. The maximum viability obtained will be in the range of 50%. If there are too many blood cells in the preparation, layer the cell suspension on top of a fresh layer of 50% Percoll and centrifuge a second time.

Follicle puncture for isolation of rat GC

This technique was originally developed using hypophysectomised, oestrogen-primed immature rats (Erickson & Hsueh, 1978). Hypophysectomy has proved to be unnecessary and expensive. Estrogen priming significantly increases the yield of GC and is equally effective in intact immature (25-day-old) rats as it is in hypophysectomised rats. There is some concern that oestrogen priming causes changes in the GC resembling atresia (Sadrkhanloo, Hofeditz & Erickson, 1987), therefore the desirability of estrogen priming should be carefully evaluated. Estrogen priming can be achieved by a single subcutaneous injection of 100 µl of 10 mg/ml diethylstilbestrol in corn oil. The ovaries should be collected 2–3 days after injection. The technique is also applicable to unprimed rat ovaries.

Protocol

1 Asphyxiate the rats by CO_2 inhalation and remove the ovaries. Place the freshly collected ovaries in ice cold H-199B.

2 Transfer the ovaries to the lid of a petri dish containing enough H-199B to cover the bottom of the lid. With the aid of a dissecting microscope remove the bursa, oviduct, uterus and fat that may be adhering to the ovary.

3 Transfer the cleaned ovaries to a fresh petri dish containing fresh medium. Attach a new disposable needle to a tuberculin syringe. The gauge of the needle is not important and the syringe acts as a convenient handle. Puncture all of the visible follicles with the needle. When the follicles are punctured, GC will begin to spill out of the follicles into the medium. Take care not to disperse the GC into the medium. Collection of the GC is much more efficient if they are concentrated. After the follicles have been punctured, gently press on the ovary to express as many GC as possible. Too much force will express the cells rapidly and more of the cells will die than is necessary.

4 Using a sterile Pasteur pipette aspirate the GC and collect in a sterile 15 ml conical centrifuge tube on ice.

5 Discard the punctured ovary, replace with another ovary and repeat the puncturing until all of the ovaries have been punctured. Add additional H-199B as necessary.

6 Centrifuge the GC at 250 g for 5 min at 4 °C and resuspend the pellet in 1 ml of tissue culture medium.

7 Count the cells in a haemocytometer using trypan blue exclusion to determine viability. Expect a yield of about 1.5×10^6 viable GC per pair of ovaries with approximately 50% viability of the GC.

Theca cells

There are two principal methods for isolation of TIC. Microdissection will work with any follicle of sufficient size for the manipulations to be performed, and the TIC obtained will be well differentiated. As with the GC, care must be taken to account for the stage of follicle development and the state of health/atresia of the follicles from which the TIC are isolated. A specialised method has been developed for obtaining relatively undifferentiated TIC from hypophysectomised rats using a Percoll density gradient purification (Magoffin & Erickson, 1988; Magoffin, 1989). It is important to use hypophysectomised immature rats with this method to obtain highly functional TIC. The method as presented does not work well with adult rats, nor has it been validated in other species. Since estrogens are potent inhibitors of ovarian androgen production, estrogen-primed rats are also not a good source of TIC.

Materials

H-199 (HEPES buffered Medium-199): Medium-199 with Hank's salts, with L-glutamine, with 25 mM HEPES buffer (Gibco, cat. no. 12350-039)

H-199B: Dissolve 100 mg of bovine serum albumin in 100 ml H-199. Prepare fresh.

CDS (Collagenase/DNase solution)

  0.4 g collagenase (Worthington, CLS 1)

  1.0 g crystalline BSA (Gibco, cat. no. 810-1019)

  0.5 ml deoxyribonuclease: DNase I, 2 mg/ml in H-199 (Sigma, cat. no. D-4527)

  50 ml H-199

Aliquot (2 ml) and store at −20 °C.

Percoll (1.055 g/ml)
113 ml Percoll (Pharmacia, cat. no. 17-0891-01)
183 ml H-199B
2 ml 100× Penicillin-Streptomycin solution (Gibco, cat. no. 600-5070)
2 ml 100× L-glutamine solution (Gibco, cat. no. 320-5030)

Equipment
Hydrometer (Fisher Scientific, cat. no. 11-555G)
Hydrometer jar (Fisher Scientific, cat. no. 08-530K)

Protocol
1 Taking care not to introduce bubbles into the solution, mix the ingredients thoroughly then pour into a sterile hydrometer jar and measure the specific gravity at room temperature with a sterile hydrometer.
2 Make sure the hydrometer is not touching the sides of the jar when a reading is made. This formula should be close to 1.055 g/ml.
3 Add small volumes (generally <1 ml) of H-199B to decrease, or Percoll to increase the specific gravity to exactly 1.055 g/ml. This solution must be prepared from sterile stock solutions using aseptic technique because Percoll solutions cannot be sterilised by filtration. Autoclaving will alter the specific gravity of the solution.
4 Store at 4 °C for up to 1 year. Upon long-term storage a small amount of evaporation will occur causing the solution to increase in specific gravity. Add H-199B as necessary to maintain the yield of TIC in the proper range.

Microdissection of TIC

Protocol
1 The ovaries should be processed through step 8 of the microdissection protocol for collecting GC. Even if the GC are not desired, it is important to remove them from the follicle wall (Fig. 2.1) to minimise contamination of the TIC.
2 After the GC have been removed, cut the follicle in half. The theca interna is only a few cell layers thick and can be carefully dissected from the theca externa with fine forceps. The theca interna will peel away from the follicle wall as a thin membrane.
3 Place the theca shells into a sterile glass 20 ml scintillation vial containing 2 ml of CDS solution, close the vial and place into a shaking water bath at 37 °C. Agitate at approximately 80 oscillations/min for 30 min.

Any sterile glass container which can be sealed can be used to digest the theca but one with a larger diameter is preferred to permit effective agitation.

4 After incubating for 30 min, gently flush the theca tissue up and down in a sterile Pasteur pipette 20 times. Do not flush vigorously because the shearing effect will kill large numbers of cells. At this point the pieces of theca will have begun to disperse, but large pieces will remain. Cap the vial and replace in the water bath and continue the incubation with agitation at 37 °C.

5 After a total of 60 min of incubation, remove the vial from the water bath and gently flush the tissue 20 times up and down in a sterile Pasteur pipette with the tip drawn finer. To decrease the size of the orifice of the Pasteur pipette, heat a sterile 9″ flint glass Pasteur pipette in a flame about midway in the tapered region until it softens. Do not pull while in the flame because the resulting constriction will be too sharp and the pipette tends to close. Remove the softened pipette from the flame, then immediately and rapidly stretch the pipette. If performed properly, the pipette will now be approximately 30″ long with a gradual taper to a very fine orifice at the centre of the heated area. Break off the pipette adjacent to the constriction at a point where the diameter of the pipette is approximately one-third to one-half the original diameter.

6 Add an equal volume of ice cold tissue culture medium and centrifuge at 250g for 5 min. Resuspend the pellet in 1 ml of medium and count the number of cells in a haemocytometer using trypan blue exclusion to determine viability. The resulting population of TIC should be >95% pure and >90% viable.

Percoll gradient isolation of rat TIC

Protocol

1 Sprague–Dawley rats should be hypophysectomised at 21 days of age and sacrificed at 25–26 days of age by CO_2 inhalation. Remove the ovaries from 12–15 rats and collect in ice-cold H-199B.

2 Transfer the ovaries to the lid of a petri dish containing enough H-199B to cover the bottom of the lid. With the aid of a dissecting microscope, remove the bursa, oviduct, uterus and fat that may be adhering to the ovary. Place the dissected ovary into a petri dish on ice containing H-199B and repeat until all of the ovaries have been dissected free from the bursa, and non-ovarian tissue.

3 Remove the top of the petri dish in the ice bucket and place, inverted,

on the stage of a dissecting microscope. Pour a small amount of H-199B into the top of the petri dish. Using fine forceps, transfer the cleaned ovaries into the fresh medium. Each of the transfers serves as an opportunity to wash the tissue in sterile medium and helps to reduce contaminants. Cut each ovary into 4–6 pieces.

4 The pieces of ovary are too small to conveniently use forceps and too large to pass through the opening in a standard Pasteur pipette. Break off the tip of a sterile Pasteur pipette close to the constriction and transfer the cut pieces of ovary into a sterile 15 ml centrifuge tube.

5 Fill the tube two-thirds full of H-199B to wash the ovaries. Allow the pieces of ovary to settle to the bottom of the tube and while the debris is still floating, remove all but 2 ml of the medium.

6 Add one 2 ml aliquot of CDS solution to the tube and swirl to mix. Using a broken Pasteur pipette, transfer the contents of the tube into a sterile glass 20 ml scintillation vial, close the vial and place into a water bath at 37 °C. Agitate at approximately 80 oscillations/min for 30 min. Any sterile glass container which can be sealed can be used to digest the ovaries, but one with a larger diameter is preferred to permit effective agitation of the ovaries.

7 After 30 min, remove the ovaries and gently flush up and down through a sterile Pasteur pipette 20 times. Do not break the pipette. The pieces of ovary should be partially digested and will pass through the pipette. Do not flush vigorously because the shearing effect will kill large numbers of cells. At this point the pieces of ovarian tissue will remain large. Cap the vial, place into the water bath and continue the incubation with agitation at 37 °C.

8 Proceed as described in Section 5 of 'microdissection of TIC'.

9 Cap the vial, place back into the water bath and incubate at 37 °C, with agitation, for a further 30 min.

10 After a total of 90 min of incubation, remove the vial from the water bath and gently flush the pieces of ovary 20 times up and down through a sterile Pasteur pipette with the opening drawn to about one-third of the original diameter. After this pipetting, the pieces of ovary should be mostly single cells with a few small aggregates. Do not attempt to digest until there are only single cells. Overdigestion yields cells that perform poorly in culture, possibly due to damage to the hormone receptors.

11 Transfer the dispersed cells into a sterile 15 ml centrifuge tube and fill two-thirds full with H-199B. Centrifuge at 250 g for 5 min at 4 °C. Resuspend the cells in 1–5 ml of tissue culture medium depending on the number of rats used. Count the cells in a haemocytometer using

trypan blue exclusion to determine the concentration of the dispersed cells.

12 Prepare 10 ml of 44% Percoll by diluting 4.4 ml of stock Percoll solution with 5.6 ml of H–199B. Pipette 1 ml of 44% Percoll into each of six 12 × 75 mm sterile culture tubes.

13 Carefully layer 2 ml of 1.055 g/ml Percoll solution on top of the 44% Percoll layer taking care not to mix the solutions.

14 Mix the cell suspension and carefully layer 1 ml of the suspension on top of the 1.055 g/ml Percoll layer. A sharp interface should be present. Avoid agitating the tubes during handling. Centrifuge the six tubes at 450 g for 20 min at 4 °C. Turn off the brake on the centrifuge and let the rotor coast to a stop to prevent disturbing the gradient after centrifugation is complete.

15 After centrifugation the tubes should appear similar to Fig. 2.5. Aspirate the top layer of cells containing granulosa, debris and other cell types first. Next, collect the entire 1.055 g/ml layer by aspiration and pool the cells from three tubes into each of two 15 ml sterile centrifuge tubes. Add 6 ml of McCoy's medium to each of the two tubes and mix by inversion. Centrifuge the tubes for 5 min at 250 g and 4 °C.

16 Resuspend the pellets in a total of 1 ml of McCoy's medium. Count the cells in a haemocytometer and determine the viability by trypan blue exclusion. The viability should be in the 98+% range because dead cells will not enter the 1.055 g/ml Percoll layer.

Establishment and maintenance of cultures

Both the GC and TIC can be cultured in culture medium such as DMEM, DMEM/F-12, and McCoy's 5a medium. Addition of serum is optional, depending on the design of the experiments. We routinely use serum-free McCoy's 5a medium supplemented with 100 U/ml penicillin, 100 μg/ml streptomycin sulphate and 2 mM L-glutamine with excellent results. Dilute the cells to the appropriate concentration and plate as soon as possible. For experiments in which steroidogenesis is the endpoint, we routinely culture 20 000 viable cells in 0.2 ml of medium in 96-well microtest plates (Falcon, cat. no. 3072), changing the medium and treatments every two days. Up to 100 000 viable cells can be cultured in 96-well plates and larger numbers of cells can be cultured in larger culture dishes. The cells will remain viable for weeks but will rapidly lose hormone responsiveness if treatments are not initiated. TIC will lose their responses within approximately 48 h after plating

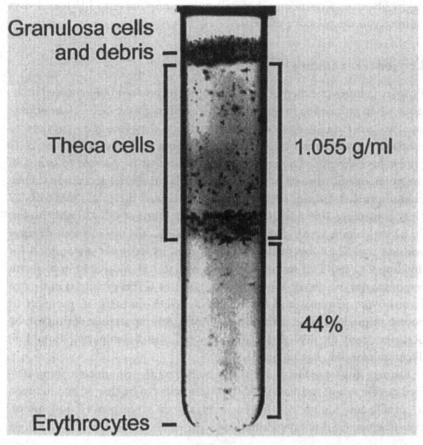

Fig. 2.5. Percoll gradient purification of theca–interstitial cells from the ovaries of hypophysectomised immature rats. The gradient shown was prepared and centrifuged as described in the protocol. This is the typical appearance of the gradients immediately after centrifugation. The granulosa cells, dead cells and various connective tissue cells do not enter the 1.055 g/ml Percoll layer and accumulate at the top of the gradient. If too many cells are loaded on the gradient, theca–interstitial cells will also be trapped here. The theca–interstitial cells accumulate at the interface of the 1.055 g/ml and 44% Percoll layers as well as floating throughout the 1.055 g/ml Percoll layer. It is not important if the interface has few cells and there are many cells floating in the 1.055 g/ml layer. The erythrocytes pellet at the bottom of the tube. A difference in the density of the 1.055 g/ml layer of 0.001–0.002 g/ml can make a large difference on the number of theca–interstitial cells that enter the gradient. Expect 15–18% recovery of the total number of cells loaded onto the gradient as purified theca–interstitial cells.

in the absence of treatments, but will remain responsive to LH for 14 days or longer in the presence of LH.

Longer-term culture

Primary cultures of either GC or TIC do not retain their differentiated functions when induced to proliferate *in vitro*. Although the cells can be stimulated to grow using serum or growth factors, they lose their capacity to produce steroids within a couple of generations as a rule. McAllister *et al.* (1989) have developed a medium formulation that is reported to maintain the steroidogenic capacity of human theca cells for several generations. They culture primary human TIC on fibronectin-coated dishes in DMEM:F-12 (1:1) containing 10% fetal bovine serum, 2% Ultroser G™, 20 nM insulin, 20 nM selenium, 1mM vitamin E, and antibiotics for three to four passages without significant loss of steroidogenic capacity. It is not uncommon for dividing GC and TIC to lose responsiveness to LH and FSH but to retain responsiveness to cAMP and forskolin. Ultroser G™ is produced in France from bovine placentas and because of a USDA embargo in products of bovine origin is not sold in the United States. An importation permit can be obtained from the USDA and Ultroser G™ can be imported from Life Technologies, Inc, Paisley, Scotland.

Primary isolates of GC and TIC can be frozen in medium containing 10% fetal bovine serum and 10% DMSO. These cells can be thawed and induced to proliferate. Unless the frozen cells are to be expanded in culture, cryopreservation is not recommended.

Troubleshooting

Obtaining granulosa cells is a straightforward process. If the viability of the cells obtained by follicle puncture is very low (<50% viable), it is likely that too much force is being used to express the cells from the punctured follicles. In granulosa preparations where the cells are scraped from the follicle wall, the scraping process can damage many of the cells if done too vigorously. The granulosa cells should exhibit a significant increase in estrogen production when stimulated by FSH in the presence of androstenedione (10^{-6} M). When culturing human cells, it is important to use human hormones, since human cells do not respond well to hormones isolated from many species. Also, be aware of the endogenous hormonal milieu from which the cells were obtained. If there was substantial endogenous stimulation of the cells, the background steroidogenesis will be high and often further stimulation with gonadotropins is impossible. It is sometimes possible to obtain low background and observe a decent FSH response by allowing the cells to incubate for 24 h in culture without hormones.

When performing thecal cell purification using Percoll gradients, it is important to determine the percentage of the total cell population that is recovered in the 1.055 g/ml layer. Generally 12–16% of the total dispersed cells applied to the gradient should be recovered. If the recovery is greater than this, there is likely to be some degree of contamination with granulosa cells. If the recovery is less than 12%, the cells are unlikely to respond well to LH in culture. If the appropriate yield is not obtained, very fine adjustments of the density of the Percoll solution should be made and the cells repurified. Hormone treatments should be initiated immediately upon plating the cells, since hormonal responsiveness declines markedly the longer the cells are allowed to incubate in culture without stimulation.

References

Erickson, G.F. & Hsueh, A.J.W. (1978). Stimulation of aromatase activity by follicle stimulating hormone in rat granulosa cells *in vivo* and *in vitro*. *Endocrinology*, 102, 1275–82.

McAllister, J.M., Kerin, J.F.P., Trant, J.M., Estabrook, R.W., Mason, J.I., Waterman, M.R. & Simpson, E.R. (1989). Regulation of cholesterol side-chain cleavage and 17α-hydroxylase/lyase activities in proliferating human theca interna cells in long term monolayer culture. *Endocrinology*, 125, 1959–66.

Magoffin, D.A. (1989). Evidence that luteinising hormone-stimulated differentiation of purified ovarian thecal-interstitial cells is mediated by both type I and type II adenosine 3′,5′-monophosphate-dependent protein kinases. *Endocrinology*, 125, 1464–73.

Magoffin, D.A. & Erickson, G.F. (1988). Purification of ovarian theca-interstitial cells by density gradient centrifugation. *Endocrinology*, 122, 2345–7.

Sadrkhanloo, R., Hofeditz, C. & Erickson, G.F. (1987). Evidence for widespread atresia in the hypophysectomised estrogen-treated rat. *Endocrinology*, 120, 146–55.

3

Anterior pituitary cells

Seon H. Shin and John V. Milligan

Introduction

The anterior pituitary gland secretes at least six major protein hormones; growth hormone (GH), prolactin (PRL), luteinising hormone (LH), follicle-stimulating hormone (FSH), thyroid-stimulating hormone (TSH) and adrenocorticotrophic hormone (ACTH). Each of these peptide hormones is secreted from a specific cell type, and is synthesised and stored in secretory granules within their respective cell types. The granules are released instantaneously by exocytosis when a secretagogue stimulates the cells.

The six different types of pituitary cell show characteristic shapes and sizes of granules when they are examined by transmission electron microscopy. Although the hormones in these granules can be identified by immunocytochemical staining with specific antibodies, the cell types of the anterior pituitary gland are still commonly classified as either acidophils or basophils. This classical method of identification is based on the staining properties of acidic and basic dyes.

Different disciplines in the physiological sciences traditionally use different animal species for research purposes. For example, most cardiovascular research is based on experiments with dogs, while cats are the main species used in neuroscience. In endocrine studies, rats have been used extensively and will continue to be used because there is such a large accumulation of baseline data that is valuable for reference purposes. For this reason, we refer only to rat pituitary tissue in the methods described herein.

The pituitary gland is located at the base of the brain. In humans, it is surrounded by a bony structure, the sella turcica, but the rat pituitary lies on a flat bone surface. The central, paler part is the neurohypophysis or posterior pituitary, while the reddish periphery is the adenohypophysis. The major part of the adenohypophysis is composed of the pars distalis, commonly termed the anterior pituitary. Below the neurohypophysis, the adenohypophysis of the pituitary gland is thinner than the peripheral region, so that the shape of the rat anterior pituitary is flat and resembles that of a butterfly.

Tissues vs. dispersed cells

The rat anterior pituitary has two symmetrical lobes. The gland can easily be cut into two equal portions (pituitary halves), one of which is used for the control and the other for the test or experimental study. Rat pituitary halves were used extensively from the mid-1960s to the early 1970s. Pituitary cell dispersion methods were then introduced and culture techniques for the dispersed cells have evolved since the early 1970s. Primary cultured cells have subsequently become very popular for testing the effects of stimulating or inhibiting agents for a number of reasons. Firstly, the problem of diffusion distances and diffusion barriers present in pituitary halves is eliminated. *In situ*, each cell in the pituitary has direct contact with a capillary blood vessel to supply oxygen and nutrients and to pick up released hormones and metabolites. However, there is no capillary flow in excised pituitary tissues. Thus, when pituitary tissues are incubated with a labelled amino acid, the amino acid only penetrates several outer layers of cells. The cells in the centre may die during the incubation period due to the lack of oxygen and nutrient supply (Farquhar, Skutelsky & Hopkins, 1975). Any hormones synthesised by the inner layers of pituitary tissue are not released immediately into the incubation medium due to diffusion barriers. Dead or asphyxiated cells in the inner layers of the tissue will slowly leak stored hormones in an uncontrolled manner, which constitutes a major source of experimental error. The second reason for the increasing use of primary pituitary cultures is that experiments with primary cultured cells require fewer pituitary glands than do those using pituitary halves, and approximately two million cells are recovered from each rat pituitary after cell dispersion. Only about 100 000 cells are usually needed for each test in a monolayer culture. Therefore, one pituitary gland will provide enough cells for 20–24 tests. At least 50 glands would be needed if similar testing were to be performed with pituitary halves. The reproducibility for hormone secretion studies is much better with primary cultured cells than with pituitary halves. This improvement is mostly due to the direct contact of each cell with a uniform concentration of secretagogue. The equal number of cells in each test group also gives a much more precise response than incubations using pooled pituitary halves (usually two or three pituitary halves for each test point). Finally, the sensitivity of the secretory response to stimulatory or inhibitory agents is much greater with cultured cells than with pituitary halves. However, the mechanism(s) of the improved sensitivity remains to be experimentally established.

Primary cultures vs. cell lines

The hormone release from anterior pituitary cells may be quantified by comparing release from the untreated control cultures with that from the treated experimental cultures. The basal rate of hormone release (i.e. from the control cultures) is commonly used as the reference value (100%), and the stimulatory and inhibitory actions are defined relative to the basal rate of hormone secretion. In most cases, other hormones in the medium do not interfere with assay of the hormone under investigation, so mixed populations of primary cultured anterior pituitary cells do not create problems for studying the rate of synthesis or secretion of a specific single hormone. A primary culture is more likely to mimic physiological conditions than an established cell line. However, as primary cultures must be prepared from fresh pituitary tissue for every experiment, this increases the cost compared to established cell lines, while also requiring more effort.

Methods of cell isolation and culture

Pituitary gland collection

Animals
Young mature male rats (Sprague–Dawley, 250 – 310 g)

Chemicals, reagents and other supplies
Diethyl ether (anaesthetic grade)
Sheet of cotton
Large glass jar with lid (5 litres)
70% alcohol
Dispersion medium (refer to cell dispersion section)
Petri dish (100 × 15 mm; Falcon Plastic, Oxnard, CA, USA)

Equipment and tools
Guillotine (Small animal decapitator, Harvard Apparatus, South Natick, MA, USA)

Sterile hood (SterilGard™ Hood, Baker, Stanford, ME, USA)
Surgical tools: coarse scissors, fine iris scissors, jeweller's forceps, surgical toe
    nail clipper, scalpel

Collection of rat pituitary glands

Protocol

1 Lightly anaesthetise individual rats by placing them in a 5-litre glass jar
   containing a cotton sheet, soaked with ether, on the bottom.
2 Remove the head quickly with the guillotine and soak in 70% alcohol for
   2 min. It should be noted that, if the decapitated heads are not wetted, too
   many hairs and particles will fly around. It is therefore essential to soak the
   decapitated head in 70% alcohol to minimise the possibility of contaminating the primary culture.
3 Remove the skin using the coarse scissors, dip the skinned head in 70%
   alcohol and take to a sterile hood.
4 Inside the hood, open the top of the skull with the surgical toe nail clipper,
   gently lift the forebrain with forceps, cut the optic nerves with fine scissors and remove the brain. The entire pituitary should be left *in situ*,
   covered by a transparent membrane. Figure 3.1 illustrates the appearance
   of the dissection at this stage of the procedure.
5 Tearing the middle part of the membrane with a sharp pair of forceps
   exposes the gland. Remove the posterior pituitary (pars nervosa), on the
   top and central part of the gland, by picking it off with the forceps; this
   also removes the intermediate lobe (pars intermedia). The anterior pituitary (pars distalis or adenohypophysis) (Fig. 3.1) is kept in place by the
   membrane.
6 When the lateral parts of the membrane are removed, free the anterior
   pituitary and pick it up gently with forceps.

Pituitary cell dispersion

The adenohypophyses are dispersed under sterile conditions using the
combination of trypsin digestion and mechanical shear forces.

Chemicals, reagents and other supplies

Trypsin (bovine pancreas type III) (Sigma Chemical Co., St. Louis, Mo.,
    USA).
Lima bean trypsin inhibitor (LBTI; type II-L) (Sigma Chemical Co.)
Bovine serum albumin (BSA; Cohn Fraction V) (Sigma Chemical Co.)

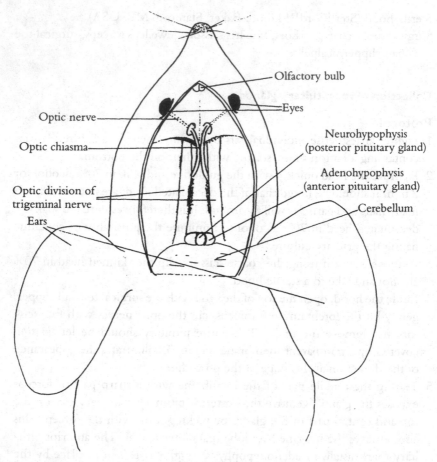

Fig. 3.1. Diagram of a rat dissection to expose the anterior pituitary gland. The skull is cut open and the cerebral cortex is removed, exposing the base of the brain. In addition to some nerves, the pituitary gland is the only tissue found in the exposed area. The gland is covered by a thin membrane, and its inferior boundary aligns to the ear of the rat (A. Chang & S.H. Shin, unpublished data).

Spinner's minimum essential medium (S-MEM, cat. no. F-14 Gibco, Grand Island, NY, USA)

Dispersion medium: S-MEM containing 0.05% trypsin and 0.1% BSA (Sigma Chemical Co.)

Petri dishes (35 × 10 mm and 100 × 15 mm, Falcon Plastic, Oxnard, CA, USA)

Glass Pasteur pipettes

Scalpel

Equipment and tools
Water bath (MagiWhirl, Blue M Electric Co., Blue Island, IL, USA)
Spinner's flask (25 ml capacity, Bellco, Vineland, NJ, USA)
Water-driven magnetic stirrer (K-791000, Kontes, NJ, USA) Benchtop
    Centrifuge (Medifuge, Baxter Scientific)
Microscope (Zeiss, Germany)
Inverted microscope (Zeiss, Germany)
Haemocytometer (Brightline, American Optical Co., Buffalo, NY, USA)

Protocol
1 Ten anterior pituitaries may be dispersed at one time. After each gland is removed, place it in one drop of dispersion medium on a Petri dish (100 × 15 mm).
2 Cut the gland into small pieces (<1 mm³) with a scalpel blade, and transfer these to a Spinner's flask containing 10 ml of dispersion medium. Stir the pituitary pieces in the medium at 37 °C in a water bath using a rotating magnetic bar suspended in the Spinner's flask. The bar should be arranged to spin at 100 rpm by a water-driven magnetic stirrer. Dispersion is aided by intermittently sucking in and out the pieces of anterior pituitary tissues suspended in S-MEM with a 9″ Pasteur pipette (between 70 and 100 times in total). Dispersion should be continued until no fragments of pituitary tissues are visible, or for 2 h.
3 Following dispersion, compact the cells into a soft pellet by centrifugation (750g for 5 min). Discard the dispersion medium and resuspend the cell pellet with the Pasteur pipette in 10 ml of S-MEM containing 0.04% LBTI.
4 After a second centrifugation, resuspend the cell pellet in 10 ml of S-MEM to wash out the residual LBTI. Following this, centrifuge the cells again and resuspend in the culture medium (refer to section 'Cell culture systems').

The anterior pituitary of a 250–300 g male rat weighs about 10 mg. Cell recovery is normally $1.5 \times 10^6 - 2.5 \times 10^6$ cells per pituitary. The recovery rates are approximately 50% of total pituitary cells (2×10^6 cells out of $4 \times 10^6 - 5 \times 10^6$ cells). In excess of 90% of the dispersed cells should be alive as judged by the trypan blue dye exclusion test. Other enzymes such as collagenase can be used to disperse cells, but trypsin digestion does not differ from collagenase treatment, as determined by cell recovery and cell sensitivity to secretagogues after a few days of culture. The S-MEM as originally formulated for suspension culture does not contain Ca^{2+} as a constituent. A Ca^{2+}-free medium helps cell dispersion and reduces the possibility of reaggregation of the cells. Therefore, the efficiency of cell dispersion is enhanced.

The 0.1% BSA added to the dispersion medium is believed to moderate the action of trypsin on cell membrane proteins and thus increases cell recovery. During the 2 h digestion period, connective tissues are softened by trypsin digestion, but the major tissue structure remains intact. Therefore, a gentle mechanical shear from the Pasteur pipette action may be used to finish the cell dispersion.

The DNA that originates from broken cells may create a problem in the formation of a compact pellet of the dispersed cells by centrifugation. Threads of DNA look like loose cotton. Since the DNA will disappear during the LBTI treatment, one can either ignore the DNA thread or, alternatively, a small amount of DNase can be added to the dispersion medium.

Cell culture systems

Monolayer cultures

Chemicals, reagents and other supplies
Carbon dioxide
Dulbecco's Modified Eagle's Medium (DMEM, Gibco)
Fetal calf serum (FCS), and horse serum (Gibco)
Sodium bicarbonate (2.7 g/l) (J.T. Baker Chemical Co., Phillipsburg, NJ, USA)
Culture medium: DMEM supplemented with 2.5% (v/v) FCS, 15% (v/v) horse serum, sodium bicarbonate (2.7 g/l), 100 U penicillin-G and 100 μg/ml streptomycin.
Test medium or Perifusion medium: DMEM containing 0.1% BSA (DMEM-BSA).
Tissue culture cluster (4 × 6 wells; Falcon 24-well Multiwell; Falcon Plastic).
Pasteur pipettes
Sample cups (Sarstedt, Numbrecht, Germany)

Equipment and tools
Water-jacketed incubator (Forma Scientific, Marietta, Ohio, USA)
Eppendorf pipette (1000 μl adjustable pipette, Hamburg, Germany)
Plastic squeeze bottle, sterile (500 ml size)

Hormone release studies using static monolayer cultures

Cell counting and resuspension

Protocol

1 After the acutely dispersed cells have been resuspended in the culture medium, estimate the cell density of the suspension by counting the number of cells in a representative aliquot, using a haemocytometer.

2 Dilute the cell suspension with additional medium to make 10^5 cells/ml and distribute 1 ml aliquots of the cell suspension into each well of a tissue culture plate (4 × 6 wells). Place these in a water-jacketed incubator at 37 °C under a water-saturated atmosphere of 5% CO_2 and 95% air. The cells are cultured in cluster plates for 2–5 days before hormone secretion is investigated.

Many red blood cells will be found in the acute dispersed cell suspension, and should not be included in the pituitary cell counts. Pituitary cells have variable sizes and shapes and are larger than red blood cells. The smaller red blood cells have a uniform size and a common concave shape. At least 50 pituitary cells should be counted in adjacent sectors of the haemocytometer to keep the counting error below 15%.

The pituitary cells continue to grow for many days but the characteristics of their hormone secretion pattern change after about 10 days. For example, the basal rate of prolactin secretion from cells maintained for longer than 10 days in culture is very low and the cells do not respond well to secretagogues.

Changing culture media

After 2–5 days, the old culture medium should be discarded by turning the tissue culture clusters upside down and shaking out the residual medium. The adherent cell monolayer is then washed once with about 1 ml of the prewarmed test medium. New medium is gently poured into the wells using a sterile plastic squeeze bottle, the sharp tip of which has been blunted by cutting it away. If the tip is too sharp, the medium may flow in a jet and cause cells to be detached from the monolayer.

Testing secretagogues

The test medium should be prewarmed and equilibrated with CO_2. A range of secretagogue doses (0 (control), 10^{-9}, 10^{-8}, 10^{-7}, 10^{-6}, 10^{-5} mol/l) is prepared with the test medium and added to each well with an Eppendorf pipette using tips that have been blunted with a scalpel blade. Quadruplicates of six different concentrations are used. The range of five log doses covers a wide range of physiological and pharmacological actions and, if a compound does not have any effect over this dose range, its physiological significance should be questioned.

The tissue culture clusters are incubated for 1, 2 or 3 h at 37 °C. After incubation, the clusters are gently shaken by swirling them a few times and 0.5–0.7 ml of sample is collected from each well into individual sample cups using Pasteur pipettes. Unless care is taken to avoid touching the bottom of a well with the tip of the pipette, cells from the monolayer on the bottom of the well may be inadevertently sucked up. If the samples cannot be assayed within 24 h, they may be stored at −20 °C .

Suspension culture / perifusion system

Chemicals and reagents

Culture medium and Perifusion medium (refer to the Monolayer culture section)

Glass wool (Corning Glass Works, Corning, NY, USA)

Syringe needle (18 gauge, Becton Dickinson & Co. Franklin Lakes, NJ, USA)

3 ml disposable plastic syringes (Becton Dickinson & Co.)

Bio-Gel P-2 (Bio-Rad Laboratories, Richmond, CA, USA):
    The Bio-Gel is swollen for at least 24 h in distilled water before use. The swollen Bio-Gel is equilibrated with perifusion medium before it is packed into a syringe.

Sample cups (Sarstedt)

Petri dishes (35 × 10 mm, Falcon Plastic)

Pasteur pipettes

Tubing (Tygon tubing, 1/32″ internal diameter, 1/16″ outer diameter and 1/16″ internal diameter, 1/8″ outer diameter, Norton, Akron, Ohio, USA)

Equipment and tools

Water bath (Blue M Electric Co.)

Peristaltic pump (Desaga peristaltic pump, Heidelberg, Germany)

Fraction collector (LKB 7000 Ultrarac, Sweden)

Four-way valve (SRV4, Pharmacia, Uppsala, Sweden)

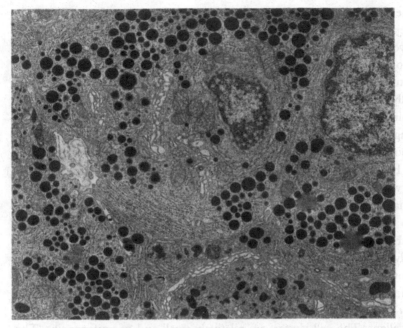

Fig. 3.2. Electron micrograph of a lactotroph grown as part of a cell cluster in a Petri dish. Lactotrophs have mature secretory granules throughout the cytoplasm and numerous immature granules. The ultrastructure of these cells is similar to lactotrophs within the intact pituitary gland *in vivo*, and they respond well to TRH stimulation in perifusion experiments.

Dynamic perifusion experiments

Incubation

Protocol

1 Suspend the dispersed cells from ten rat pituitaries in 20 ml of culture medium (approximately 2×10^6 cells in 2 ml) after they have been washed with medium containing LBTI, and then wash with fresh medium (refer to the Monolayer culture section).

2 Distribute aliquots of the cell suspension (2 ml) to ten Petri dishes, place in a water-jacketed incubator at 37 °C under a water-saturated atmosphere of 5% CO_2 and 95% air and culture for 2–5 days (Fig. 3.2).

It is important to note that the cells must be cultured in Petri dishes in this procedure. In contrast to the so-called 'tissue culture' dish, which is positively charged to promote monolayer formation, the Petri dishes are not treated, so the dispersed cells preferentially attach to each other to form cell clusters (Joneja *et al.*, 1993). The cell clusters float in the medium and are easy to recover for use in a perifusion column or any other experiments where monolayers are undesirable.

Pituitary cells that are attached to the bottom of a dish as a monolayer are usually lifted by treating them with trypsin for a few minutes or by physically scraping them off with a rubber policeman. The trypsin treatment may damage receptors and other membrane proteins, and may therefore distort the normal response pattern. Furthermore, removal of cells from the mono-layer by scraping may also lead to physical damage and, while the damaged cells will reseal themselves, it is not known whether resealed cells exhibit normal responses.

To avoid these potential problems, positively charged beads such as DEAE-Sephadex can be used as a substrate for cell attachment. The cells readily anchor to the DEAE-Sephadex beads and are thus easily collected, without major disruption, for a perifusion study. However, cell clusters are simpler to handle. The ultrastructure of clustered cells does not differ from *in situ* pituitary cells (Joneja *et al.*, 1993), and their response to secretagogues is excellent. The cell clusters are relatively small and of uniform size. The cluster sizes are apparently self-regulating and we see no obvious problem of a diffusion barrier. Therefore we believe that the cell clusters are better for perifusion experiments than cells recovered from a monolayer culture.

Loading the perifusion chamber

Protocol

1 Place a suspension of clustered pituitary cells between two layers of 0.3 ml Bio-Gel P-2 in the barrel of a 3 ml disposable plastic syringe. The barrel outlet should first be plugged with a 3 mm layer of glass wool.
2 Attach tubing sufficient to reach the fraction collector from the outlet, and clamp.
3 Filter the Bio-Gel P-2 beads, suspended in perifusion medium, through the glass wool until 0.3 ml of Bio-Gel has accumulated. Then filter the suspension of cell clusters through the Bio-Gel layer. Finally, load 0.3 ml of Bio-Gel beads on top of the cell clusters; leave about 0.8 ml of perifusion medium on top of the second Bio-Gel layer.
4 Push an 18-gauge syringe needle through the top of the rubber part of the plunger from the disposable syringe. Cut away the plastic syringe adapter hub of the needle, and connect the 18-gauge metal tubing to plastic infusion tubing that is open-ended.

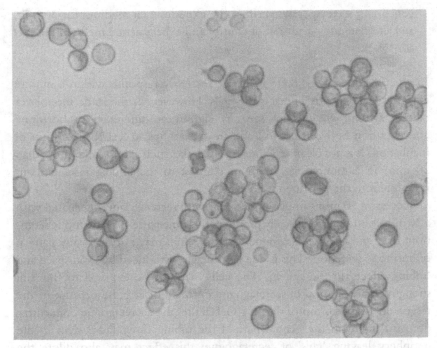

Fig. 3.3. Rat anterior pituitary cells photographed under phase-contrast microscopy (magnification × 503). The cells remain spherical in shape for the first 48 h in culture, after which time they extend cytoplasmic processes (M. Evans & S.H. Shin, unpublished data).

5 Push the needle fitted plunger into the syringe barrel until 0.5 ml perifusion medium is left on top of the Bio-Gel layer. Air and medium in the syringe should be vented out of the open-ended tubing before the tubing is connected to the outflow tubing of the pump.

6 Prime the pump with perifusion medium. Place the syringe, packed with Bio-Gel and cell clusters, with its attached tubing, in a 38 °C water bath and open the clamp on the outflow tube.

7 Place the perifusion medium reservoir in the water bath. Perifuse the cell clusters by pumping medium through the syringe at a rate of 0.5 ml/min. After an hour, the basal rate of prolactin secretion is stable and this period is chosen for the pre-experimental perifusion period.

8 During the experimental period, collect successive 1.5 ml volumes of perifusates in disposable cups every 3 min. A 30 min perifusion period (or collection of 10 fractions) is recommended to establish the initial control level of secretion.

9 A four-way valve (SRV-4) is used to switch control medium to medium

containing secretagogue; perifuse the experimental medium for 30 min, and alternate the perifusion of control and experimental medium (Shin *et al.*, 1992).

In a perifusion system, there is a large secretory response within a minute after the secretagogue contacts the cells. However, in the static monolayer culture system, we do not see any significant difference in hormone concentration between control and treated groups in less than 30 min of incubation. It is not clear why the response takes such a long time (more than 30 min) in a monolayer system in contrast to the almost instantaneous response from the perifused cell clusters.

Successive 30 min stimulations of cells with secretagogue, alternated with a 30 min control period usually produces the same pattern of hormone secretion. Brief pulses of secretagogue for 1 to 3 min may not show any pattern of hormone secretion. This is due to artefacts introduced by dead space and 'tubing effects' in the system. The cells may not be exposed to the 'full strength' concentration of secretagogue during a brief pulse because (i) the reserve medium on top of the Bio-Gel dilutes the secretagogue concentration, (ii) liquid in the central part of the tubing flows faster than in the periphery leaving 'tails' of secretagogue; this effect may also dilute the released hormones, reducing the concentration in the outflow, and (iii) the 3 min sampling time cannot accurately resolve short 'spikes' of hormone release or distinguish between release that occurs on different magnitudes of time and scale. For example, when thyrotropin-releasing hormone (TRH) is perifused there is an initial surge of prolactin release, believed to be triggered by an increased cytosolic calcium concentration ($[Ca^{2+}]_i$). After the surge, steady but much lower rates of release, mainly stimulated by diacylglycerol (DAG) (Gershengorn, 1985) can be seen.

The monolayer culture system is an excellent method for screening a large number of samples and to examine dose–response relationships. The perifusion system, in contrast, provides far more detailed and useful information than the monolayer system, but it generates a large number of perifusate samples for assay. The monolayer system can measure only the final cumulative release of hormone, whereas the perifusion system measures moment-to-moment changes in the rate of hormone secretion. Typically, adenohypophysial hormones are secreted in an initial burst when a secretagogue first contacts the cells. Subsequently, secretion either continues at a lower rate or else ceases, even though perifusion with the same concentration of secretagogue is maintained.

Purification and enrichment of individual pituitary cell types

Primary cultured anterior pituitary cells are a mixed population of six major cell types and many subtypes. This mixed population has both advantages and disadvantages. If only hormone release is being studied, the sensitivity and specificity of current immunoassay procedures is such that hormones of interest can be characterised without interference from the remaining hormones. The specificity of a secretagogue on pituitary cells can be easily determined by assaying for all six pituitary hormones in a single sample. However, when paracrine effects are being determined (i.e. where the activity of one cell type affects adjacent cells) or when signal transduction systems are investigated, the mixed population is not satisfactory. All six different types of pituitary cells share the same signal transduction systems and it is difficult to clarify which cell type is responsible for changing the activity or level of a common signal transduction component such as cyclic AMP, inositol trisphosphate (IP_3) or $[Ca^{2+}]_i$. Individual transduction systems cannot therefore be studied in a quantitative manner in a mixed population of cells.

Experimentation with single cells after identification of a particular cell type is feasible for studies of cytosolic calcium and membrane potentials. However, adenylyl cyclase and phospholipase C systems cannot be studied because of the analytical problems for single cells. Purification (actually enrichment) of a particular cell type in a population to 80–90% 'purity' allows them to be considered as practically homogeneous. Enriched cell preparations prepared by discontinuous gradient centrifugation have thus been used to define the role of cyclic AMP and cyclic GMP on GH secretion (Lussier et al., 1991). The enrichment technique for somatotrophs is simple enough to be practical.

Chemicals, reagents and other supplies

Phosphate buffered saline (PBS: 0.01 M NaH_2PO_4 plus 0.15 M NaCl, adjust to pH 7.4 with NaOH)

Bovine serum albumin (Sigma Chemical Co.)

Centrifuge tubes

Equipment and tools

Centrifuge (IEC B-29A Centrifuge, Damon/IEC Division, Needham Heights, MA, USA)

Plastic disc on a wire
Specific gravity hydrometer (Fisher Scientific)
Hand refractometer (for specific gravity measurement) (N-1, Atago, Japan)

Protocol

1 Form the first discontinuous gradient by layering BSA solution in PBS (specific gravity; 1.068 g/cm^3, upper layer) onto a BSA solution in PBS (specific gravity; 1.087 g/cm^3, lower layer) in a centrifuge tube.
2 When either the top gradient solution or the cell suspension is loaded, place a wire-mounted plastic disc just on top of the lower layer, and carefully run the solution onto the surface of the plastic disc with a Pasteur pipette. Gently remove the plastic disc without mixing the solutions.
3 Disperse ten rat pituitaries, suspend in 1 ml of 1% BSA in PBS and then gently load onto the top layer of the gradient. Centrifuge for 30 min at 3500 rpm (1000 g).
4 Collect the cell layer located at the interface between the two density layers, after the top layer has been carefully removed with a Pasteur pipette. Resuspend the collected cells in 10 ml of 1% BSA in PBS and concentrate into a pellet by centrifugation at 1500 rpm (200 g).
5 After they are resuspended in 1 ml of 1% BSA in PBS, reapply the cells to a second discontinuous gradient (1.071 g/cm^3, upper layer and 1.087 g/cm^3, lower layer) and again centrifuge at 1000 g for 30 min. A cell population, of which 80–90% are GH-secreting somatotrophs, will be found at the interface between the two density layers (Snyder & Hymer, 1975).
6 Resuspend the cells in 10 ml of 1% BSA in PBS and recover them by centrifugation. All of the procedures should be performed at 4 °C.

Several different techniques have been used in attempts to purify a specific pituitary cell type. Isotonic density gradients with low viscosity can be made using Percoll, a high molecular weight silica complex coated with PVP. Centrifugation can be performed at room temperature with low g forces using this material. However, enrichment of the mixed population of cells in the pituitary will not be successful unless one cell type is already predominant. Although the proportion of lactotrophs or thyrotrophs can be selectively increased by manipulation of the endocrine state of the animal, GH secreting cells (somatotrophs) which constitute 40% of the cells in a normal rat pituitary, are the only pituitary cell type that has been enriched sufficiently, with any success, for extensive study.

Measurement of cytosolic calcium concentration $[Ca^{2+}]_i$ in anterior pituitary cells

Measurements of $[Ca^{2+}]_i$ are critical to the understanding of pituitary hormone secretion. Fluorescent dyes such as Fura-2 , Indo-1 and Fluo-3 are used for $[Ca^{2+}]_i$ measurements. The sensitivity of the method of measurement is such that changes of $[Ca^{2+}]_i$ can be determined in a single cell. If the cell type can be identified, these changes can be related to changes in function. Individual cells are usually identified with a reverse haemolytic plaque assay provided they continuously secrete hormones without stimulation; that is, they have a small but constant basal secretion. Lactotrophs, for example, secrete prolactin continuously. A mixture of dispersed pituitary cells and protein-A coated red blood cells are cultured together until a monolayer forms. The monolayer is washed and covered with new medium containing prolactin antiserum and guinea pig complement. The prolactin (antigen) secreted from lactotrophs reacts with prolactin antibody and in the presence of complement, the antigen-antibody reaction causes lysis of the red blood cells. Plaques, proportional to the amount of secreted prolactin are formed by the lysis (Neill & Frawley, 1983). A pituitary cell, surrounded by a plaque, can thus be reliably identified as a lactotroph. Although the positively identified cells can be readily examined for changes of $[Ca^{2+}]_i$, there are no reliable techniques yet available to measure changes in the components of either the adenylyl cyclase or the phospholipase C systems in individual cells.

The non-polar form of a fluorescent dye such as Indo-1, AM is loaded into the primary cultured cells and its non-polar ester is hydrolysed by cellular esterases. The hydrolysed Indo-1 becomes a polar compound and remains inside the cells. When Indo-1 is complexed with Ca^{2+}, the emission wavelength of its fluorescence changes from 490 nm (free Indo-1) to 405 nm (Indo-1-Ca^{2+} complex) when an excitation wavelength of 355 nm is used. The emission intensity at 405 nm is proportional to the $[Ca^{2+}]_i$ concentration. When $[Ca^{2+}]_i$ is increased, the emission intensity increases at 405 nm and decreases at 490 nm. Using the ratio of the emission intensities at 405 nm and 490 nm increases the sensitivity of the measurement of changes in $[Ca^{2+}]_i$ because when $[Ca^{2+}]_i$ is increased, the intensity at 405 nm increases reciprocally to the decrease at 490 nm. Another advantage is that the ratios are independent of total dye concentration. Ionised Indo-1 leaks slowly out of the cells, and thus the basal level of emission intensity at 405 nm slowly decreases. However, when the ratios are used, the baseline remains constant, even though the intracellular concentration of ionised Indo-1 slowly decreases due to leakage. A single peak dye such as Fluo-3 can be used for the measurement of $[Ca^{2+}]_i$ when a laboratory is not equipped with a fluorometer that can handle the ratios. The sensitivities of emission intensities with the single peak dye are usually higher than the ratiometric dyes. The emission intensity of Fluo-3 at 525 nm is proportionally increased with $[Ca^{2+}]_i$ when an excitation wavelength of 490 nm is used.

Although the fluorometric measurement of $[Ca^{2+}]_i$ in single pituitary cells is a sensitive and well-established technique, there are a number of pitfalls. For example, in many single-cell studies, dynamic changes of $[Ca^{2+}]_i$ have been measured at room temperature, which represents a substantial deviation from physiological body temperature. Another issue that is often overlooked is that of the protein content of the medium. In primary cultured pituitary cells, the basal hormone secretion is low and the cells do not respond well to secretagogues in the absence of protein (usually BSA) from the medium. Many investigators therefore use a 0.1% BSA supplement in the medium for hormone secretion studies. However, protein-free medium is still commonly used for single-cell studies since protein creates problems with the micro-pipettes used for electrophysiological studies.

Pituitary cell perifusion for simultaneous analysis of $[Ca^{2+}]_i$ and hormone release

Numerous investigators have tried to correlate $[Ca^{2+}]_i$ measurements with hormone secretion without actually making direct measurements. However, with the set-up used for $[Ca^{2+}]_i$ measurements in single cells, simultaneous measurements of $[Ca^{2+}]_i$ and hormone release are difficult. In an attempt to undertake such simultaneous measurement, we have used a perifusion system using a fluorometric cuvette to hold the cells in suspension (Shin et al., 1993).

Chemicals, reagents and supplies
Dulbecco's Modified Eagle's Medium without phenol red (Gibco)
Dimethyl sulphoxide (DMSO, Baker Chemical Co., Phillipsburg, NJ, USA)
Acetoxymethyl Indo-1 (Indo-1, AM, Molecular Probes, Inc., Eugene, OR, USA)
Bovine serum albumin (Sigma Chemical Co.)
Sephadex G-150 (Pharmacia); swollen for more than 48 h in distilled water
Manganese chloride (Mallinckrodt)
Ionomycin (Sigma Chemical Co.)
Triton X-100 (Sigma Chemical Co.)

Silicone tubing (1/23″ internal diameter, 5/64″ outer diameter, New Brunswick Scientific, New Brunswick, NJ, USA)

Glass wool (Corning Glass Works)

Syringe needle (18 gauge, Becton Dickinson and Co.)

00-size rubber stopper

Tubing (Tygon R-3603, 1/32″ internal diameter, 3/32″ outer diameter, Norton)

Sample cups (24 × 14 mm, Sarstedt)

Equipment and tools

Semi-micro size disposable cuvette (Dynalon Disposable polystyrene cuvette, 1.5 ml, 10 × 4 × 45 mm, Canlab Scientific Products, Canada)

Drill (Moto-tool kit, Dremel, Racine, WI, USA)

Benchtop centrifuge (Medifuge, Baxter Scientific)

LS-50 Spectrofluorometer (Perkin-Elmer Ltd., Buckinghamshire, UK)

Peristaltic pump (12000 Vario Perpex, LKB)

Jacketed column (Kontes Glass Co., Vineland, NJ, USA)

Circulation pump (PolyTemp, Polyscience Corporation, Niles, IL, USA)

$[Ca^{2+}]_i$ measurement

Construction of a cuvette for perifusion

Protocol

1 A semi-micro size disposable cuvette is used to reduce the dead space of the Sephadex gel matrix. Drill the lower right corner of the cuvette face which receives excitation spectra, and insert silicone tubing (1/23″ internal diameter, 5/64″ outer diameter) into the hole. Plug the lower part of the cuvette with glass wool to prevent leakage.

2 Pack the cuvette with a Sephadex G-150 gel matrix equilibrated with DMEM-BSA (Medium without phenol red is used for these fluorometric studies because phenol red interferes with fluorometry).

3 Cap the cuvette with a 00-size rubber stopper pierced by 18 gauge steel tubing (a syringe needle), then connect the steel tubing to the medium reservoir with the Tygon tubing.

Loading the cells with Indo-1 and placing them in the cuvette

Cultured cells, 2–5 days old, are collected from the suspension by centrifugation (750 g for 10 min) and resuspended in 10 ml of DMEM. Ten microlitres of dimethyl sulphoxide (DMSO) solution containing 10 nmole Indo-1, AM is added to the 10 ml DMEM. (The final concentration of Indo-1, AM is 10 μmol/l.) After incubation for 30 min at 37 °C, the cells are recovered by centrifugation (750 g for 10 min) and resuspended in 10 ml DMEM-BSA. The suspension of Indo-1 loaded cells is left for 30 min at room temperature. The cells are then centrifuged and resuspended in 50 μl of DMEM-BSA. The cuvette containing 1 ml of Sephadex G-150 is equilibrated with perifusion medium. An additional 0.2 ml of DMEM-BSA is added to the top of the Sephadex matrix. The cell suspension is then injected into the Sephadex G-150 matrix using a Pasteur pipette. The injection site of the cells loaded with Indo-1 is a position between 9 and 18 mm from the bottom of the cuvette, which is where the excitation light path projects from the spectrofluorometer. The cells loaded with Indo-1 are left for at least 1 h at room temperature to allow sufficient time for hydrolysis of the ester of Indo-1, AM inside the cells.

Perifusion

Two jacketed columns are used to warm both the control and the secretagogue-containing media to 37 °C before either is pumped through the cuvette. A jacket surrounding the cuvette chamber in the spectrofluorometer and the jacketed columns are connected in series with the silicone tubing. Water at 39 °C is pumped through the tubing with the circulation pump. The perifusion medium is drawn through the cuvette with a peristaltic pump at a rate of 0.4 ml/min for a period of 30 min before experiments are performed so the cells can adjust to a perifusion environment at 37 °C. During the experimental period, the flow rate is 0.4 ml/min and the perifusate is collected in sample cups at the rate of one sample per min. The cells are perifused with perifusion medium or medium containing an appropriate concentration of secretagogues. The dead space between the medium reservoir and the cuvette, and between the cuvette and the sample cup is 0.3 ml and 0.9 ml, respectively.

Fluorometry

For Indo-1, the excitation wavelength is 329 nm, and the emission wavelength is 405 nm. Excitation and emission slit widths are set to 5 nm and 10 nm, respectively. The relationship between intensity of fluorescence and $[Ca^{2+}]_i$ is determined using the ionomycin and Mn^{2+} quenching technique. The $[Ca^{2+}]_i$ is then determined by the method of Tsien $et\ al$ (1982) with

$K_d = 250$ nM. The ionomycin solution is diluted in DMEM or DMEM-BSA solution to make the appropriate concentrations. A stock solution of 1 mol/l manganese chloride in distilled water is diluted in DMEM or DMEM-BSA to make 0.1 mmol/l Mn^{2+} to quench extracellular free Ca^{2+}. Figure 3.4 illustrates the results of a perifusion of lactotrophs sequentially exposed to TRH, ionomycin, Mn^{2+} and Mn^{2+}+ionomycin, with simultaneous measurement of $[Ca^{2+}]_i$ and prolactin release.

Ionomycin is often used for calibration of basal $[Ca^{2+}]_i$. When protein is present, ionomycin binds to protein reducing its availability to bind to the cells. Under such conditions, the ionomycin does not function fully. BSA should therefore be omitted from the media when ionomycin is used to calibrate the system for $[Ca^{2+}]_i$. Furthermore, a matrix must be present in the cuvette for a successful perifusion. The matrix fixes the cells in position inside the cuvette. Products commonly used are either Sephadex gels or Bio-gels. Sephadex G-150 interferes less with light transmission than Sephadex G-25 and is not compressed during the perifusion period.

Troubleshooting

It is common practice to keep primary cultures of pituitary cells for 2–5 days in culture before experimentation. During this time, no medium change is necessary. The rate of hormone secretion in response to secretagogues is greater from primary cultured pituitary cells after 2–5 days in culture than the comparable release from acutely dispersed cells or non-dispersed pieces of pituitary tissue. It is commonly believed that the lower level of hormone release from acutely dispersed cells is due to damage to receptors by the trypsin treatment. However, there is no direct experimental evidence showing that membrane-bound active receptors are actually damaged during cell dispersion and subsequently recover during the culture period to respond fully to stimulation by receptor agonists. The increased response only lasts for about 1 week. After about 10 days, the sensitivity to secretagogues and the basal hormone secretion both begin to decrease, although the cells continue to grow well. The hormone secretion pattern changes completely after 2 weeks but the cells still appear healthy. Thus, a healthy morphological appearance and a normal growth rate are no indication that the cells are in the same condition as they are in the intact gland. This is why most investigators make measurements of cell function within 2–5 days after the beginning of the primary culture. Primary cultures of pituitary cells cannot be sustained by subculturing because the characteristics of the hormone secretion pattern change over time.

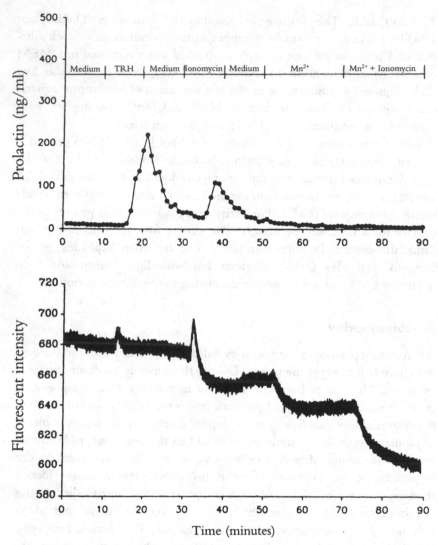

Fig. 3.4. Simultaneous measurements of $[Ca^{2+}]_i$ and prolactin release from lactotrophs in a perifusion system. Perifusion was performed in the following sequence: medium, thyrotrophin-releasing hormone (TRH) (10^{-6} mol/l), medium, ionomycin (5×10^{-6} mol/l), medium, Mn^{2+} (10^{-3} mol/l), and Mn^{2+} (10^{-3} mol/l)+ionomycin (5×10^{-6} mol/l). The media were supplemented with 0.1% bovine serum albumin for the entire perifusion period. Fluorescence intensity was recorded while perifusate was collected, throughout the experiment. The prolactin level in the perifusate was measured by radioimmunoassay. (Reprinted by permission from Shin et al., 1993, Elsevier Science Inc.)

Culture medium

The ideal culture medium and conditions for anterior pituitary cells would sustain all normal physiological functions including hormone synthesis and release. However, commonly used culture conditions are not ideal, and there is a tendency to follow precedents without critical evaluation when selecting a medium for a particular cell type. Poorly defined biological products such as fetal calf serum (2.5%) and horse serum (15%) are often used as supplements for defined media such as DMEM or Ham's F-12 nutrient mixture. These sera are sources of protein, unknown essential nutrients, hormones and growth factors. Although serum supplementation is an absolute requirement for robust cell growth in culture, it is not known whether a supplement of 2.5% fetal calf serum and 15% horse serum provides the optimal conditions for pituitary cell growth.

Minor differences in the constituents of different culture media can sometimes generate major differences in the responses of cultured cells. For example, the presence of ascorbic acid and other reducing agents can cause significant differences in prolactin secretion. Dopamine, believed to be the prolactin-release inhibiting factor (PIF) which regulates basal physiological prolactin release, is readily oxidised in solution by molecular oxygen (O_2) in the air. Dulbecco's MEM contains neither ascorbic acid nor glutathione. However, M-199 contains small amounts of ascorbic acid (0.05 mg/l or 0.28 μmol/l) and glutathione (0.05 mg/l or 0.16 μmol/l) which, as reducing agents, increase the half-life of dopamine in solution. During a 180 min incubation period in M-199 at 37 °C, the dopamine concentration falls by only 20%, in contrast to a 50% reduction after a 36 min incubation (T $1/2 = 36$ min) of dopamine in DMEM.

It is interesting to note a historical coincidence. M-199 was the most commonly used incubation medium in the 1960s when modern endocrinology was budding. If investigators had used DMEM at this time, they would never have been able to identify dopamine as the PIF because the dopamine concentrations needed to inhibit prolactin release in a DMEM incubation are too high to be considered 'physiological'. Dopamine concentrations below 100 nmol/l are unable to inhibit prolactin release in the DMEM medium (without ascorbic acid). Physiological concentrations in the hypophysial portal blood are only a few nmol/l. In a M-199 incubation, prolactin release is inhibited by 1 nmol/l dopamine plus 0.1 mmol/l ascorbic acid. Ascorbic acid not only chemically protects dopamine from oxidation but also potentiates dopamine action as the PIF. The ascorbic acid content of plasma is relatively high (23–85 μmol/l) and is believed to serve as a physiological supplement to the action of dopamine as the PIF.

The estrogenic effect of phenol red

Most commercial culture media, such as DMEM and M-199 contain phenol red, which is used to visualise changes of pH in the culture medium and it has been assumed to be biologically inert. However, it is in fact a mildly estrogenic compound. Such estrogenic action is a significant factor for lactotrophs and other pituitary cells because estrogen serves as a powerful stimulator of prolactin synthesis and secretion, and has a strong negative feedback effect on gonadotrophin secretion. Estrogen also up-regulates somatostatin receptors in lactotrophs (Cooper & Shin, 1981). Male lactotrophs *in situ* do not respond to the inhibitory action of somatostatin on prolactin release, but somatostatin is a powerful inhibitor of prolactin release in estrogen-primed male rats or in female rats. Primary cultures of male pituitary cells respond well to the inhibitory action of somatostatin, in contrast to poor responses of male pituitary tissues. The estrogenic action of phenol red and the presence of estrogens in horse serum are likely to enhance this inhibitory response by the induction of new somatostatin receptors.

References

Cooper, G.R. & Shin, S.H. (1981). Somatostatin inhibits prolactin secretion in the estradiol primed male rat. *Can. J. Physiol. Pharmacol.*, **59**, 1082–8.

Farquhar, M.G., Skutelsky, E.H. & Hopkins, C.R. (1975). Structure and function of the anterior pituitary and dispersed cells. *In vitro* studies. In *Ultrastructure in Biological Systems Volume 7. The Anterior Pituitary*, ed. A. Tixier-Vidal & M.G. Farquhar, pp. 83–135. New York: Academic Press, Inc.

Gershengorn, M.C. (1985). Thyrotropin-releasing hormone action: mechanisms of calcium-mediated stimulation of prolactin secretion. *Rec. Prog. Horm. Res.*, **41**, 607–53.

Joneja, M., Reifel, C.W., Murphy, M.L. & Shin, S.H. (1993). Ultrastructural changes in rat mammotropes following incubation with dopamine. *Experientia*, **49**, 836–9.

Lussier, B.T. French, M.B., Moor, B.C. & Kraicer, J. (1991). Free intracellular Ca^{2+} concentration ($[Ca^{2+}]_i$) and growth hormone release from purified rat somatotrophs I. GH-releasing factor induced Ca^{2+} influx raises $[Ca^{2+}]_i$. *Endocrinology*, **128**, 570–82.

Neill, J.D. & Frawley, S. (1983). Detection of hormone release from individual cells in mixed populations using a reverse hemolytic plaque assay. *Endocrinology*, **112**, 1135–7.

Shin, S.H., McAssey, K., Heisler, R.L. & Szabo, M.S. (1992). Phenoxybenzamine selectively and irreversibly inactivates dopaminergic D_2-receptors on primary cultured rat lactotrophs. *Neuroendocrinology*, **56**, 397–406.

Shin, S.H., Soukup, C. Pang, S.C., Kubiseski, T.J. & Flynn, T.G. (1993). Measurement of prolactin release and cytosolic calcium in estradiol-primed lactotrophs. *Life Sci.*, **53**, 1605–16.

Snyder, G. & Hymer, W.C. (1975). A short method for the isolation of somatotrophs from the rat pituitary gland. *Endocrinology*, **96**, 792–6.

Tsien, R.Y., Pozzan, T. & Rink, T.J. (1982). Calcium homeostasis in intact lymphocytes: cytosolic free calcium monitored by a new intracellular trapped fluorescent indicator. *J. Cell Biol.*, **94**, 440–5.

4

Pancreatic β-cells

Leonard Best

Introduction

The pancreatic islets of Langerhans represent approximately 1% of the bulk of the mammalian pancreas. The number of islets varies between a few hundred and several thousand, depending on the species. Between 200 and 400 islets can be routinely isolated from a single rat pancreas. The anatomy of the endocrine pancreas has been reviewed extensively (Kloppel & Lenzen, 1984). Briefly, each islet consists of approximately 1000–2000 cells, comprising a number of different cell types. In most cases, 70–80% of these are insulin-secreting β-cells. The remainder consists of glucagon-secreting α-cells, somatostatin-secreting D-cells and cells which secrete pancreatic polypeptide. These different cell types appear to be arranged in a specific manner, with the β-cells occupying the inner 'core' of the islet and the other cell types located peripherally. This may be an important feature for the overall functioning of the islet, and there is evidence that islet cells may be chemically and electrically coupled, thus forming a functional syncitium. Each islet has cholinergic and adrenergic nerve endings and a vascular supply, the latter perfusing the β-cell core prior to the other cell types. These points should always be borne in mind when interpreting data obtained from single isolated cells in terms of intact islet function *in vivo* or *in vitro*. It is the purpose of this chapter to consider techniques for the preparation and culture of functional β-cells together with potential research applications.

Applications of β-cell culture

Normal β-cell function

The most important physiological stimuli for insulin release are nutrients (particularly glucose, but also some amino acids), gastrointestinal hormones and certain neurotransmitters. It is beyond the scope of this chapter to review our current understanding of β-cell biology. However, the introduction of increasingly precise and sensitive techniques for the study of β-cell biochemistry, physiology and molecular biology has provided an enormous amount of information concerning the biochemical, electrical and biophysical events, and their interactions, in regulating function both in the normal and diabetic β-cell.

The development of methods for the isolation of intact, functional pancreatic islets has enabled extensive studies of β-cell physiology and biochemistry including uptake and metabolism of nutrients (e.g. glucose and amino acids), hormone and neurotransmitter actions and intracellular signalling, ionic fluxes (e.g. $^{45}Ca^{2+}$, $^{86}Rb^+$) associated with changes in membrane potential and electrical activity, phosphorylation of cellular proteins and insulin secretory activity. It has also permitted the investigation of the pharmacology of these processes, notably the mechanisms of action of drugs such as sulphonylureas which exert profound stimulatory effects on insulin release and are widely used in the treatment of non-insulin dependent diabetes mellitus (NIDDM).

Studies employing intact pancreatic islets have, however, always been complicated by the presence in intact islets of non-β-cells. More recent techniques for the dispersal of islets into single cells, and their subsequent primary culture, now permit the study of single functional β-cells. This preparation is highly suited for techniques such as fluorescence microscopy and imaging and patch-clamp (Fig. 4.1). From the use of such techniques, we now have a considerable understanding of the regulation of, for example, cytosolic $[Ca^{2+}]$ and the importance of this process in mediating insulin release. We also have greater insight into the function of numerous ion channels in the β-cell and their role in modulating electrical and secretory activity. It is now possible to study the process of exocytosis in a single β-cell using the patch-clamp technique to measure cell capacitance, an estimate of the surface area of the plasma membrane. The application of molecular biological techniques to β-cell research has also led to the molecular identification of numerous proteins thought to be of critical importance in β-cell function, such as K_{ATP} channels, glucose transporters and glucose phosphorylating enzymes.

The purification of β-cells on a relatively large scale can now be achieved by fluorescence-activated cell sorting (FACS) on the basis of glucose-induced changes in autofluorescence (for example, see Van De Winkel & Pipeleers, 1983). This technique is particularly suitable for studies which require β-cells in large numbers and for the study of β-cell populations and heterogeneity.

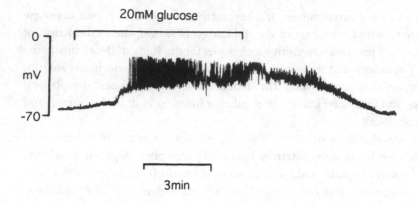

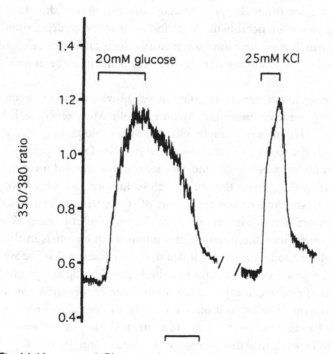

Fig. 4.1. Upper panel: Glucose-induced depolarisation and electrical activity in a single rat ß-cell, recorded using the 'perforated patch' configuration of the patch-clamp technique. Lower panel: Changes in cytosolic [Ca^{2+}], assessed by 350/380 fluorescence ratio, in response to glucose and a depolarising concentration of K^+ in a single rat ß-cell loaded with the Ca^{2+}-sensitive dye Fura-2.

The vast majority of studies of pancreatic islet cell function has utilised rodent tissue, predominantly rat and mouse. A significant recent development has been the increasing availability of human islet tissue (Ricordi *et al.*, 1988). Since there are clear species differences with respect to β-cell ion channels, transporters, etc., it will be of vital importance to characterise human β-cells fully as a prerequisite for investigating β-cell abnormalities associated with diabetes mellitus.

β-cell function in diabetes mellitus

Most cases of diabetes mellitus arise as a result of an absolute or relative lack of insulin released from the β-cells in response to stimuli. Thus, a major aspect of β-cell research will continue to be the study of the molecular defects underlying different types of diabetes mellitus, and the development of an increasingly rational approach towards the treatment of this group of diseases. These studies have utilised several animal models of diabetes mellitus, such as the *ob/ob* and *db/db* mouse and the GK rat (models of NIDDM) and the NOD mouse and BB rat (models of IDDM). The advent of human islet isolation offers the additional prospect of investigations into β-cell function in human forms of the disease. Several molecular abnormalities have already been identified in β-cells in certain forms of NIDDM (Poitout & Robertson, 1996).

Pancreatic islet/β-cell transplantation

The preparation and culture of pancreatic β-cells, in addition to permitting studies of β-cell physiology and pathology, is also beginning to show clear potential for the eventual treatment of some types of diabetes mellitus by β-cell transplantation (Sutherland *et al.*, 1996). Clearly, optimisation of islet cell preparation and their function and survival following implantation will be important factors leading to progress in this area (Clayton & London, 1996).

β-cell morphology and characterisation

The preparative methods described below have been used extensively in this laboratory for the isolation and culture of pancreatic β-cells from the rat and to a lesser extent, though with equal success, from the mouse. Similar techniques are also, in principle, applicable to the preparation and culture of islet cells from other mammalian species, including humans.

Since pancreatic β-cells are invariably prepared from isolated islets, it is

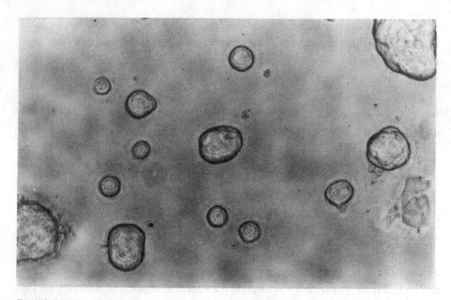

Fig. 4.2. Light micrograph of rat pancreatic ß-cells prepared by exposure to Ca^{2+}-free medium with 0.025% (w/v) trypsin ($\times$ 400 magnification).

most important that the β-cells can be distinguished from other types of islet cell. As noted above, the non-β-cells are considerably less numerous than the β-cell population, comprising at most 30% of the total islet cell number. This proportion appears to be reduced even further during islet preparation and dispersal, possibly because non-β-cells are more fragile or because, through occupying a peripheral location in the islet, these cells are more susceptible to enzymatic and mechanical damage. Thus, in our experience, the average β-cell preparation will comprise at least 80% β-cells. The ultrastructural morphometry of the β-cell has been described in detail elsewhere (Dean, 1973). However, even under bright field illumination, β-cells have a characteristic appearance. At a magnification of $\times$300–400, β-cells from both rat and mouse appear circular or sometimes slightly ovoid with a smooth, well-defined plasma membrane and a granular cytosol (Fig. 4.2). The nucleus can often be discerned as a paler area in the cell. Freshly prepared rat β-cells range from 12 to 16 μm in diameter, corresponding to a volume of approximately 1500 μm^3. This is approximately two to three times the size of non-β-cells (Pipeleers et al., 1985). The progressive culture of rat β-cells in a supplemented culture medium often results in the gradual appearance of cells of considerably larger diameter (up to 50 μm; Fig. 4.2 and Mathers et al., 1985). Whether this phenomenon results from amitotic growth of a single cell or the fusion of several normal size β-cells is unclear.

A number of pancreatic exocrine cells are also likely to be found in an islet cell preparation. These are typically of a comparable size to β-cells but can be distinguished from them by their irregular pattern of dark zymogen granules.

A positive identification of β-cells in a mixed islet cell culture can be achieved with greatest certainty on the basis of the functional characteristics of the β-cell. For example, the haemolytic plaque assay (see Chapter 3) has been used to identify β-cells on the basis of their insulin secretory activity in response to glucose (Bosco et al., 1995). Similarly, in cases where another aspect of β-cell function, such as electrical activity or cytosolic $[Ca^{2+}]$, is under study, the responsiveness of the cell to a raised concentration of glucose (8–20 mM) or to a hypoglycaemic sulphonylurea may be used to positively identify β-cells (Fig. 4.1). In the latter case, tolbutamide is the most convenient of the hypoglycaemic sulphonylurea drugs to use, since its effects, unlike those of glibenclamide, are completely reversible.

β-cell heterogeneity

It has been suggested that β-cells exist as discrete populations, on the basis of their insulin secretory response to glucose and other secretagogues (Pipeleers, 1992). More specifically, it has been suggested that β-cells show intercellular differences with respect to glucose responsiveness. A concentration-dependent recruitment is thought to occur in terms of both glucose-induced insulin release and biosynthesis, whilst some β-cells appear to be unresponsive even to maximal concentrations of glucose. Whilst we have not attempted a formal assessment of heterogeneity in preparations of rat β-cells, we have rarely encountered cells which fail to show an electrical response to a maximal or near-maximal concentration of glucose (16–20 mM). When considering the possibility and extent of β-cell heterogeneity, it should be borne in mind that islet cells undergo a variable amount of enzymatic and mechanical cell damage during the preparative process. The rate and extent of recovery from this damage during subsequent culture of the cells is also likely to vary.

Source of cells

The majority of studies of β-cell function involves the use of pancreatic islets isolated from rodents, most commonly rats or mice. The first method described for isolation of intact islets involved their microdissection. This method has the benefit of avoiding the use of digestive enzymes, but is only applicable to species such as the mouse, where the islets are clearly visible under a binocular microscope. Furthermore, microdissection is a laborious and time-consuming process. This technique has been effectively superseded by a method of releasing islets from the surrounding exocrine tissue by digestion of the latter using collagenase (Lacy & Kostianovsky, 1967). As with most tissue preparative techniques employing collagenase, the choice of enzyme and the digestion conditions are crucial. The following procedures have been used routinely in this laboratory for the isolation of intact islets from rats (and occasionally from mice) and for the subsequent dispersal of islets into functional β-cells.

Materials

Standard incubation medium: NaCl (135 mM), KCl (5 mM), MgSO$_4$ (1 mM), NaH$_2$PO$_4$ (1 mM), CaCl$_2$ (1.2 mM), glucose (4 mM) and HEPES (10 mM); pH adjusted to 7.4 with NaOH.

Ca^{2+}-free medium. Consists of the above but with 2 mM MgSO$_4$, no added Ca^{2+} and 1 mM EGTA.

Collagenase; *Clostridium histolyticum*, class 4 (Worthington Biochemicals, Freehold, NJ, USA) or type 'P' (Boehringer Mannheim, Germany).

Trypsin; Bovine (Worthington, Freehold, NJ, USA).

Culture medium; RPMI 1640 buffered with 25 mM HEPES (Gibco BRL, Paisley, Scotland, UK).

Penicillin-streptomycin (both 5000 U/ml; Gibco BRL).

Fetal calf serum (Gibco BRL).

Preparation of rat pancreatic β-cells

Protocol

In order to minimise the risk of microbial contamination, all dissecting instruments should be sterilised by washing in 70% ethanol. All solutions are filtered via a 0.2 μm filter. The risk of infection can be further reduced by carrying out as much of the procedure as possible in a laminar flow cabinet or similar sterile environment.

1 Sacrifice the animal by stunning and cervical dislocation, swab the abdomen with ethanol and open the abdominal cavity.

2 Clamp the pancreatic duct as closely as possible to the duodenum. Using fine 'butterfly' scissors, make a small incision in the pancreatic duct at a higher point, just distal to the junction with the bile duct.

3 Inject approximately 20 ml incubation medium into the duct via a fine

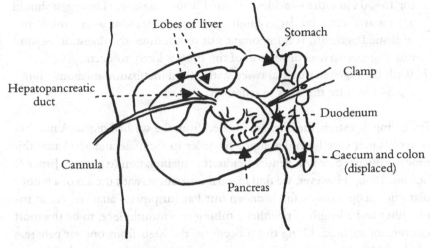

Fig. 4.3. The position and appearance of the rat pancreas *in situ* following inflation with medium via the hepato-pancreatic duct.

Portex™ catheter, thus inflating the pancreas. Figure 4.3 illustrates the anatomical position and appearance of a rat pancreas following inflation in this manner. The organ should then be separated from the duodenum and spleen. The prior inflation with medium assists in removal of the pancreas, mincing and subsequent digestion with collagenase. In the case of the mouse pancreas, the organ can first be removed, then inflated using a fine syringe needle.

4 Wash the pancreas in medium in a sterile petri dish, trim away the fat and lymph nodes, then transfer to a sterile 50 ml polythene tube in which the mincing, digestion and washing procedures can be carried out.

5 Mince the pancreas with scissors for 1–2 min and wash with incubation medium to remove endogenous digestive enzymes. Remove the excess medium by aspiration with a Pasteur pipette and add the collagenase. For the preparation of rat islets, we find that 7–8 mg collagenase/pancreas (equivalent to approximately 5 ml of mince) is optimal, although this amount may vary slightly depending on the specific activity of collagenase and other proteases in the enzyme preparation.

6 Incubate the tissue at 37 °C in a shaking water bath at approximately 120 cycles/min. The required duration of the incubation may also vary depending upon the activity and composition of the collagenase. However, with the majority of preparations, 20–25 min will be sufficient. At this point, it is normally necessary to shake the tube vigorously by hand

for 15–30 s in order to achieve optimal tissue digestion. The digest should appear as a dense but largely homogeneous suspension with a minimum of tissue fragments. It is important not to continue the digestion beyond this stage, otherwise disruption of the islets is likely to occur.

7 Wash the digested material twice with 40 ml incubation medium, allowing 3–4 min for the islets to sediment.

Following digestion, the islets are isolated from the washed digest. A number of techniques have been developed in order to facilitate and accelerate this process, usually based on density gradient centrifugation (e.g. Lake, James & Sutton, 1986). However, we find manual separation, with the aid of a binocular microscope using a finely drawn out Pasteur pipette attached via an in-line filter and a length of polythene tubing to a mouthpiece, to be the most convenient method. Using this procedure, the islets from one rat pancreas can be isolated under virtually aseptic conditions in approximately 30 min. The islets are initially transferred to a petri dish containing fresh sterile medium, and subsequently re-picked into a 15 ml conical centrifuge tube (Costar, Cambridge, MA, USA). The isolated islets are then gently sedimented by centrifugation ($150g$ for 5 min) and the supernatant medium aspirated.

The intact islets, obtained as described above, are then dispersed into cells. Two basic methods can be employed for dispersal. A degree of islet cell dispersal can be achieved by a brief incubation (3–5 min at room temperature with gentle agitation) of the islets in a Ca^{2+}-free medium, supplemented with EGTA (see above). The islets are then centrifuged at $150\ g$ for 5 min and resuspended in approximately 1 ml RPMI culture medium ('complete', containing penicillin-streptomycin (both 50 U/ml) and fetal calf serum (5% v/v)). Partial dispersal can be achieved by gently triturating the islets 15–20 times with a 1 ml Gilson pipette tip. This method has the advantage that the cells have a minimal exposure to digestive enzymes. However, islet cell dispersal using this method is invariably incomplete, resulting in a large proportion of cell clusters (see Fig. 4.2).

A significantly greater yield of single β-cells can be obtained by short-term incubation with trypsin. The islets are incubated with continuous agitation in 5 ml Ca^{2+}-free medium (see above) containing 0.025% (w/v) bovine trypsin at 37 °C for approximately 5 min. In order to terminate the incubation, 5 ml RPMI complete culture medium is added. The islets/cells are sedimented by centrifugation ($150\ g$ for 5 min) and resuspended in approximately 1ml complete RPMI with gentle trituration (approximately ten passages via a 1 ml Gilson pipette tip).

Establishment and maintenance of cultures

The isolated β-cells, prepared by either of the above methods, can be plated either directly onto dishes or ethanol–washed glass coverslips if necessary. We generally use 30 mm dishes (tissue culture treated; Nunclon, Nunc, Denmark) which conveniently form the basis of an incubation chamber for subsequent experimentation. Aliquots of approximately 50 µl of cell suspension are applied to the centre of the dish and the cells incubated in an atmosphere of humidified air at 37 °C. After a period of 2 h, the majority of viable cells will have adhered to the dish, and a further 2.5 ml RPMI can be added prior to culture. The use of HEPES–buffered medium obviates the need for incubation in a CO_2-rich atmosphere, and does not appear to compromise the integrity of the cells or the responsiveness to various β-cell stimuli.

We generally find that β-cells prepared by the above methods are in optimum condition (on the basis of glucose-responsiveness) after 2–3 days of culture. When trypsin is used, an extra day or two may be required for the cells to reach optimum condition. Typically, β-cells prepared either with or without trypsin retain their integrity and function for up to 2 weeks. It may be found desirable to replace the medium after 1 week, though when culturing the cells at low density (up to a few hundred per dish), we have not found this to be necessary. Since β-cells do not appear to replicate in culture, one is restricted in practical terms to primary cultures. With the notable exception of transformed β-cell lines, continuous β-cell cultures are not normally possible. Similarly, it is not normally practicable, at least for research purposes, to store primary β-cell preparations by cryopreservation. However, this process has been employed for the preservation of pancreatic endocrine cells in large numbers for the purpose of transplantation (see, for example, Tze & Tai, 1990).

A number of studies have focused on defining the optimal conditions, in terms of composition of culture media, for the long-term survival of β-cells in primary culture (for example, see Ling, Hannaert & Pipeleers, 1994). In summary, the survival of rat β-cells appears to be optimised by the presence of metabolisable substrates (e.g. glucose; optimal at 10 mM), by fetal calf serum (possibly due in part to the presence of hormones and growth factors) and by isobutylmethylxanthine (which raises cellular levels of cyclic AMP).

Troubleshooting

The major difficulties encountered in primary culture of pancreatic β-cells are bacterial/fungal infections and cell damage. The former can be minimised by using sterile materials and instruments and by carrying out the preparative procedure in a sterile or semi-sterile environment. If frequent infections persist, it may be helpful to wash the islets several times with sterile medium prior to dispersal.

The problem of cell damage can be equally frustrating. The main points to consider are possible damage from enzymes and from mechanical agitation. Different batches of collagenase, even from the same supplier, are likely to have varying activities of collagenase itself and also other unspecified proteases. The optimal approach is to screen several batches to find the one that yields best results, and then buy a large quantity of that batch. Some suppliers (e.g. Boehringer) will provide samples of specific batches for this purpose. In order to minimise mechanical cell damage, avoid vigorous shaking and pipetting (trituration), and especially bubble formation in order to reduce the 'shearing' effect. Trituration should be gentle (approximately 1 cycle/2 s).

The 'calcium paradox'

Methods for the dispersal of intact islets into individual cells invariably involve the use of Ca^{2+}-free (and in some cases Mg^{2+}-free) media. We have consistently noted that prolonged exposure of pancreatic β-cells to Ca^{2+}-free media results in a greater extent of cell damage and hence low yields of viable cells. It is possible that this could be related to the so-called 'calcium paradox'. This phenomenon occurs in certain types of electrically excitable cells, notably those such as cardiac myocytes and β-cells which have voltage-sensitive Ca^{2+} channels (vsccs) and a Na^+/Ca^{2+} exchange system (Hearse, Baker & Humphrey, 1980; our own unpublished observations). Briefly, exposure of the cells to Ca^{2+}-free media results in Na^+ entry via vsccs. The subsequent restoration of Ca^{2+} to the medium then results in a high rate of Ca^{2+} entry by a process of 'reverse' Na^+/Ca^{2+} exchange. It is likely that supranormal levels of intracellular Ca^{2+} cause irreversible cell damage.

Our studies with Fura-2 loaded β-cells have suggested that, when Ca^{2+} is restored to the medium following prolonged exposure (5–10 min) to Ca^{2+}-free conditions, cytosolic $[Ca^{2+}]$ can rise to very high levels. Furthermore, many cells appear to be unable to recover their low resting intracellular Ca^{2+} levels, at least during short-term (<60min) measurements. It has been reported that, in myocytes, verapamil can ameliorate the calcium paradox, presumably by blocking vsccs and thus reducing Na^+ entry. In our hands, attempts to improve β-cell viabilty by the use of verapamil have produced equivocal results.

References

Bosco, D., Meda, P., Thorens, B. & Malaisse, W.J. (1995). Heterogeneous secretion of individual B cells in response to D-glucose and to non-glucidic nutrient secretagogues. *Am. J. Physiol.* **268** (3 pt. 1), C611–18.

Clayton, H.A. & London, N.J.M. (1996). Survival and function of islets during culture. *Cell Transpl.* , **5**, 1–12.

Dean, P.M. (1973). Ultrastructural morphology of the pancreatic β-cell. *Diabetologia*, **9**, 115–19.

Hearse, D.J., Baker, J.E. & Humphrey, S.M. (1980). Verapamil and the calcium paradox. *J. Mol. Cell. Cardiol.*, **12**, 733–9.

Kloppel, G & Lenzen, S. (1984). Anatomy and physiology of the endocrine pancreas. In *Pancreatic Pathology*, ed. G. Kloppel & P.U. Heitz, pp. 133–53. Edinburgh, UK: Churchill Livingstone.

Lacy, P.E. & Kostianovsky, M. (1967). A method for the isolation of intact islets of Langerhans from the rat pancreas. *Diabetes*, **16**, 35–9.

Lake, S.P., James, R.F.L. & Sutton, R. (1986). An improved isolation method for rat pancreatic islets. *Transplantation Proceedings*, **18**, 1817–18.

Ling, Z.D., Hannaert, J.C. & Pipeleers, D.G. (1994). Effects of nutrients, hormones and serum on survival of rat islet β-cells in culture. *Diabetologia*, **37**, 15–21.

Mathers, D.A., Buchen, A.M.J., Brown, J.C., Otte, S.C. & Sikora, L.K.J. (1985). Rat pancreatic islet cells in primary culture: occurrence of giant cells amenable to patch clamping. *Experientia*, **41**, 116–18.

Pipeleers, D.G. (1992). Heterogeneity in pancreatic β-cell population. *Diabetes*, **41**, 777–81.

Pipeleers, D. G., In't Veld, P.A., Van de Winkel, M., Maes, E., Schuit, F.C. & Gepts, W. (1985). A new *in vitro* model for the study of pancreatic A and B cells. *Endocrinology*, **117**, 806–16.

Poitout, V. & Robertson, R.P. (1996). An integrated view of beta-cell dysfunction in type-II diabetes. *Ann. Rev. Med.* **47**, 69–83.

Ricordi, C., Lacy, P.E., Finke, E.H., Olack, B.J. & Scharp, D.W. (1988). Automated method for isolation of human pancreatic islets. *Diabetes* **37**, 413–20.

Sutherland, D.E.R., Gores, P.F., Hering, B.J., Wahoff, D., McKeehan, D.A. & Gruessner, R.W.G. (1996). Islet transplantation – an update. *Diabetes-Metab. Rev.* **12**, 137–50.

Tze, W.J. & Tai, J. (1990). Successful banking of pancreatic endocrine cells for transplantation. *Metabolism* , **39**, 719–23.

Van De Winkel, M. & Pipeleers, D. (1983). Autofluorescence-activated cell sorting of pancreatic islet cells: purification of insulin-containing B-cells according to glucose-induced changes in cellular redox state. *Biochem. Biophys. Res. Commun.*, **114**, 835–42.

5

Adrenocortical and adrenomedullary cells

Matthias M. Weber, Christian Fottner and Dieter Engelhardt

Introduction and application

In mammals, the adrenal glands are composed of two separate endocrine tissues, an outer cortex and an inner medulla. Both components are surrounded by a common capsule but differ greatly in their developmental origin and endocrine function. The mesodermal-derived adrenal cortex secretes a large variety of steroids which are classified as glucocorticoids, mineralocorticoids and androgens. The adult mammalian adrenal cortex is composed of three morphological zones: an outer zona glomerulosa, the zona fasciculata, and an inner zona reticularis. Functionally, aldosterone is synthesised in the zona glomerulosa and regulated primarily by angiotensin II, whereas glucocorticoids and androgens are formed in the zonae fasciculata and reticularis primarily under the influence of adrenocorticotrophin (ACTH), which is the most important trophic hormone of the two inner zones. In addition to ACTH, a variety of other hormones, growth factors, and cytokines take part in the regulation of adrenocortical cell function. Therefore, the primary culture of adrenocortical cells, which can be enriched for glomerulosa or fasciculata–reticularis cells, offers a useful and well-established tissue culture model to study the regulation of metabolic pathways and enzymes involved in steroid formation and secretion. In contrast to the cortex, the adrenal medulla is a neural crest-derived endocrine organ, whose chromaffin cells synthesise, store and release the catecholamines noradrenalin and adrenalin. Adrenal medullary chromaffin cells represent highly modified postganglionic sympathetic neurons which, *in vitro,* retain their capacity to secrete catecholamines in response to acetylcholine and other agonists. Therefore, adrenal chromaffin cell cultures can readily be used for kinetic experiments with studies of excitation and secretion. Furthermore, medullary cells have been widely applied as a cell model to study the functional properties of the sympathetic nervous system.

Bovine adrenocortical cells

The availability of large amounts of bovine adrenal tissue and the relative simplicity of culturing bovine adrenal cells has made adult bovine adreno-cortical cells an excellent primary-cell culture model for investigating the regulation of adrenocortical cell growth and steroidogenic cell function. Upon stimulation, bovine adrenocortical cells synthesise and secrete various steroid hormones. Since cortisol is the major steroid secreted by bovine fas-ciculata/reticularis cells in primary culture, and ACTH is the main inducer of adrenocortical steroidogenesis, the steroidogenic response of bovine adrenocortical cells is usually assessed by assaying the accumulation of corti-sol in the medium by radioimmunoassay (RIA). Similarly, the secretion of aldosterone in response to ACTH, K^+ or angiotensin II can be assayed by RIA of the supernatant of bovine glomerulosa cells in primary culture (Penhoat et al., 1988). Alternatively, the steroidogenic response of adreno-cortical cells can be assessed by the addition of labelled precursors (e.g. [^{3}H]pregnenolone or [^{3}H]progesterone) to the cells. At the end of the incubation period, steroids are extracted, separated by thin layer and gas chromatography and the synthesised radioactive metabolites are quantified by scintillation counting. The induction of key-enzyme genes of the steroidogenic pathway can be monitored by Northern blotting of RNA, extracted from adrenocortical cell cultures before and after treatment. In addition, many growth factor receptors are expressed by bovine adrenocor-tical cells, and the regulation of their expression can be assessed by radio-ligand assay, Northern blotting and Scatchard analysis. Since adrenocortical cells grow in vitro, they can also be used to evaluate the effect of various reagents on cell growth as determined by cell counting or [^{3}H]thymidine incorporation. ACTH however, which in vivo causes cellular hypertrophy and hyperplasia of the adrenal cortex, has an antimitogenic effect on bovine, rat, human, and tumorous adrenocortical cells in culture. A possible explana-tion for this paradox is that ACTH in vivo may be an indirect mitogen which stimulates local production of growth factors that then induce adrenocorti-cal cell division by paracrine effects. When bovine adrenocortical cells are treated with ACTH or cAMP in vitro, they change their morphological appearance and the fibroblast-like cells become more rounded and retracted. A variety of hormones, neuropeptides, and growth factors have been found to be mitogenic for adrenocortical cells in vitro, including angiotensin II, fibroblast growth factor-2 (FGF-2), vasopressin, insulin, insulin-like growth factor-I (IGF-I), thrombin, transferrin, epidermal growth factor (EGF), anti-oxidant nutrients (ascorbic acid, α-tocopherol, selenium), bovine serum albumin (BSA), and fibronectin. In contrast, transforming growth factor-β (TGF-β), and prostaglandin E_1 have been reported to have an inhibitory effect on adrenal cell proliferation.

Materials and reagents

Tissue culture plasticware (Falcon GmbH, Heidelberg, Germany)

Sieves, 50 μm, 100 μm and 150 μm (Bender & Hobein, Zurich, Switzerland)

Ascorbic acid, α-tocopherol, bovine serum albumin (BSA) fraction V, BSA (fatty acid-free), collagenase II, DNase I, ethylene diamine tetra-acetate (EDTA), fibroblast growth factor-2 (FGF-2), insulin-like growth factor-I (IGF-I), insulin, low-density lipoproteins, selenious acid, thrombin, transferrin, trypsin (Sigma, Deissenhofen, Germany)

Percoll (1.13 g/ml, Pharmacia, Uppsala, Sweden)

Phosphate buffered saline (PBS), Fetal calf serum (FCS), horse serum (HS), M-199 cell culture medium with Earle's salts, amphotericin-B, L-glutamine (Biochrom, Berlin, Germany)

NaHCO$_3$, gentamicin (Merck, Darmstadt, Germany)

Dulbecco's modified Eagle's medium (DMEM), nutrient mixture F12 (F12, Gibco, Grand Island, NY, USA)

Source of bovine adrenal glands

Adrenal glands from 2–3-year-old steers are obtained from the local slaughterhouse. Immediately after the animal has been killed, the lungs, heart, oesophagus, stomach and intestines are removed and the steers are cut in half. After this procedure, only the liver, kidneys, and a large perirenal fat mass remain attached to the inner surface (Fig. 5.1). Due to its midline position, the right adrenal gland will be damaged while cutting the steers in half, and cannot be used. The left adrenal gland can be found at the cranial pole of the left kidney and is usually embedded deeply in the perirenal fat tissue. The gland should be removed under sterile conditions. After removal, each adrenal gland is submerged in sterile ice-cold PBS containing gentamicin (52 μg/ml) and amphotericin B (0.5 μg/ml). The tissue container should be placed on ice in a second box and transferred to the laboratory as soon as possible. This procedure ensures rapid lowering of the organ temperature, and allows storage and transport of the tissue without loss of cell viability.

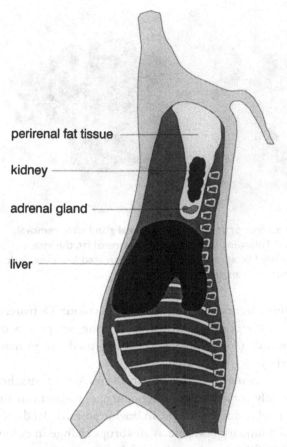

perirenal fat tissue

kidney

adrenal gland

liver

Fig. 5.1. Location of bovine adrenal glands within the body mass.

Bovine fasciculata–reticularis cell isolation

Protocol

In order to reduce the risk of contamination, only intact adrenal glands which are surrounded by periadrenal fat should be used (Fig. 5.2A).

1 Remove the pericapsular tissue of the adrenal gland, leaving the capsule intact. Wash the organ with sterile PBS and place in a dish filled with tissue culture medium. In our hands, five adrenal glands are required in order to obtain five confluent 12-well culture dishes (3.8 cm²/well).

2 Cut each adrenal gland longitudinally (Fig. 5.2B), separating the adrenocortical tissue from the medulla, and store in a fresh container filled with sterile ice-cold PBS. In contrast to the yellow colour of the adult human

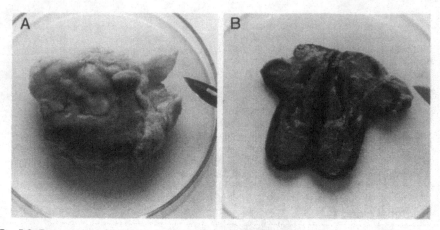

Fig. 5.2. Dissection of bovine adrenocortical tissue. A, Adrenal gland after removal from the perirenal fat tissue. B, Following removal of the perirenal fat, the intact adrenal gland is cut longitudinally. The adrenal medulla is demarcated from the dark red cortex as a pale grey tissue and can be easily scraped off.

adrenal cortex, the bovine adrenocortex is of a dark red colour. Demarcate the bovine medulla from the cortex as a pale grey tissue, scrape this off with a scalpel, and discard. (Alternatively, it can be used for primary culture of medullary cells.)

3 Separate the fasciculata–reticularis cells from the capsule and attached glomerulosa cells by gently scraping the adrenocortical tissue from the capsule with a sterile scalpel and fine forceps. In bovine adrenal glands, the glomerulosa–fasciculata boundary is visible as an abrupt change in colour from brown to red. A careful microdissection of the two zones is necessary to separate glomerulosa from fasciculata cells. Unless used for glomerulosa cell culture, discard the capsule with the attached glomerulosa cells, and combine the fasciculata–reticularis tissue fragments of all adrenal glands in a single 10 cm tissue culture dish containing a small volume of medium.

4 In order to facilitate enzymatic cell separation, cut the tissue into very small fragments and finely mince the remaining pieces with a scalpel. Transfer the tissue to a container containing 100 ml of tissue culture medium (M-199, gentamicin (52 µg/ml), and amphotericin B (0.5 µg/ml)).

5 Add freshly prepared collagenase II (1 mg/ml) and DNase I (75 µg/ml), and incubate the suspension in a gently shaking water bath at 37 °C for 45 min.

6 After digestion, filter the cells twice through a sieve with 140 µm and 100

μm mesh openings, respectively. For this purpose, pour the digested tissue onto the filter and gently stir with a sterile spoon while washing with 300 ml of tissue culture medium (at 37 °C) containing antibiotics and 10% FCS. It is important not to use mechanical force, in order to avoid crushing of the cells and consequent cytotoxic effects of lysosomal enzymes. Transfer the 300 ml cell suspension to six 50 ml Falcon tubes and centrifuge for 5 min at 400 g. Resuspend the cell pellet of each tube in 50 ml of tissue culture medium at 37 °C and pellet again for 5 min at 400 g.

7 Resuspend the cell pellet in a 50 ml centrifuge tube containing 15 ml of freshly prepared prewarmed (37 °C) iso-osmotic Percoll solution (for 100 ml of iso-osmotic Percoll solution (1.07 g/ml) combine 10 ml of 1.5 M NaCl with 49.3 ml Percoll stock solution and add 40.7 ml double-distilled water). Centrifuge the Percoll/cell suspension in a horizontal swing bucket centrifuge for 10 min at 730 g. After centrifugation, red blood cells and contaminating tissue fragments are pelleted at the bottom of the vials, whereas the fasciculata–reticularis cells are concentrated in a layer on top of the Percoll solution. Carefully aspirate the layer of fasciculata–reticularis cells with a sterile 10 ml pipette, transfer to a 50 ml Falcon tube, bring to a final volume of 50 ml with culture medium, and centrifuge for 5 min at 240 g.

8 Wash the resulting cell pellet twice in M-199 medium, in order to remove the remaining Percoll. After the last wash, carefully discard the supernatant, and resuspend the final cell pellet in a defined volume of serum containing growth medium (see below). Using this Percoll centrifugation procedure, more than 70% of the red blood cells and cell debris can be removed from the adrenocortical cell suspension.

9 Resuspend the cells in complete growth medium, and determine the cell number in an aliquot using a haemocytometer or Coulter counter. Determine the cell viability by the trypan blue exclusion test; this usually exceeds 90%. At this stage, the cells can be plated in culture dishes or frozen for long-term storage in liquid nitrogen.

Establishment and maintenance of bovine adrenocortical cell cultures

Protocol

1 Plate the cells at a density of approximately 10^6 cells/ml in prewarmed M-199 supplemented with L-glutamine (290 ng/ml), gentamicin (52 μg/ml), amphotericin B (0.5 μg/ml), 10% fetal calf serum and 5% horse serum (complete growth medium) and incubate at 37 °C in a moist atmosphere with 5% CO_2 in air. In our hands, the preparation of five

bovine adrenal glands yields approximately 6×10^7 cells. Before plating, the cell suspension must be thoroughly mixed by pipetting up and down, to obtain a homogenous cell suspension and an even distribution of cells into each well.

2 After 24 h, change the medium in order to remove cell debris and unattached cells. More than 80% of seeded cells do not attach during the first 24 h and are removed with the first change of medium. Thereafter, replenish the medium every day. Cells usually reach confluence after 72–96 h incubation in serum-containing growth medium. At this time, the cell density will be approximately 1.6×10^5 cells/well (12-well plate) and the cell viability should be 95%.

3 When the cells are used for stimulation experiments, exchange the medium for serum-free medium (M-199 supplemented with L-glutamine (290 ng/ml), gentamicin (52 μg/ml) and amphotericin B (0.5 μg/ml)) 24 h prior to the experiment, in order to avoid the potentially harmful effect of unknown substances in the serum and to reduce pseudosubstrate effects.

Pseudosubstrate effects arise when high concentrations of steroids accumulate and interact with steroidogenic enzymes. Antioxidant nutrients (ascorbic acid (100 μM), α-tocopherol (1 μM), and selenium (50 nM)), have been found to reduce pseudosubstrate effects and are therefore added to the induction medium by some groups (Hornsby & MacAllister, 1981). Similarly, metyrapone and antioxidants are required to maintain aldosterone synthesis by bovine glomerulosa cells (Crivello, Hornsby & Gill, 1982). When confluent adrenocortical cells are incubated with serum-free induction medium without any growth promoting additives, they remain growth-arrested in the G_1 phase. While this is ideal for testing induction of cell differentiation, it may not be adequate for cell proliferation assays. Therefore, the cells should be plated at a lower density (e.g. 0.5×10^6 cells/ml) in serum-reduced or growth-promoting serum-free defined medium when proliferation studies are performed. A minimal serum-free defined medium which has limited growth promoting activity should contain FGF-2 (100 ng/ml) and insulin (1 μg/ml) or IGF-I (100 ng/ml). A more sophisticated serum-free defined medium which supports continuous proliferation and passaging of adrenocortical cells, similar to serum-supplemented medium, includes the following additives: FGF-2 (4 nM), insulin (2 nM), thrombin (100 mU/ml), low density lipoproteins (10 μg/ml), transferrin (100 μg/ml), fatty acid-free BSA (500 μg/ml), ascorbic acid (100 μM), α-tocopherol (1 μM), selenium (50 nM), and antibiotics. This medium can also be used for long-term culture in combination with fibronectin-coated (2 μg/cm²) tissue culture plates (Simonian, White & Gill, 1982).

Long-term culture and passage

Primary adrenocortical cells can be maintained in the same tissue culture dish for up to 120 h while still retaining differentiated cell function and morphology. We do not use passaged cells for induction experiments since, in our experience, subcultured cells change in biochemical and functional behaviour even after one passage, and are eventually overgrown by fibroblasts. However, adult bovine adrenocortical cells can easily be subcultured and maintained in monolayer culture for up to 60 generations. For cell passaging, confluent cells are incubated for 1 to 5 min at 37 °C in a solution containing 0.05% trypsin/0.02% EDTA, in phosphate buffered saline. The trypsin is then neutralised by the addition of culture medium containing 15% fetal calf serum, after which the cells are collected by centrifugation, and plated at a 1:5 ratio in growth medium. A variety of medium formulations for bovine adrenocortical cells are described in the literature. An alternative basal medium is a 1:1 mixture of Dulbecco's modified Eagle's Medium (DMEM) with Ham's F12 medium. Although serum-free media have been described, which support the long-term proliferation of adrenocortical cells (Simonian et al., 1982), the use of serum-containing medium is still superior in promoting cell growth (Hornsby & McAllister, 1981).

Cryopreservation

Protocol

1 Resuspend the cell pellet from five adrenal glands in 8.5 ml of complete cell growth medium, then add 1 ml of FCS and 0.5 ml dimethyl sulphoxide (DMSO).
2 Mix, and transfer the cell suspension into cryopreservation tubes.
3 Store material in liquid nitrogen after continuously lowering the temperature of the vials to −70 °C.
4 To plate the cryopreserved cells in culture dishes, defrost the cells quickly while placing the tube in a water bath at 37 °C.
5 Replace the cryopreservation medium immediately with complete growth medium and plate the cell suspension into culture dishes.

Separation of bovine glomerulosa and fasciculata–reticularis cells

The protocol described above can also be used for the preparation of bovine glomerulosa cell cultures. Glomerulosa cells are characterised by aldosterone production and a very low activity of the steroidogenic enzyme 17-α-hydroxylase, whereas bovine fasciculata–reticularis cells secrete predominantly cortisol and exhibit a very strong 17-α-hydroxylase activity. To prepare crude glomerulosa cell preparations, the glomerulosa cells which remain attached to the capsule after microdissection of fasciculata–reticularis cells are completely scraped off the capsule and treated according to the protocol for fasciculata–reticularis cells. For a more standardised method of zonal separation, the cleanly dissected adrenal cortex can be cut with a microtome delivering slices of 0.5 mm, as described by Penhoat *et al.* (1988). Using this method, only the first (outer) slice is used for preparation of glomerulosa cells, whereas the two following slices are used for culturing fasciculata–reticularis cells. In contrast to the rat adrenal gland, the thick bovine capsule does not allow easy separation of glomerulosa and fasciculata cells simply by stripping. Therefore, the adrenocortical tissue adhering to the capsule will always contain contaminating cells from the inner adrenal cortex. Additionally, glomerulosa and fasciculata cells can be separated by unit gravity sedimentation as described by Crivello *et al.* (1982). In brief, a crude glomerulosa cell suspension is layered on top of a 0.3–3% BSA gradient, and allowed to sediment for 2 h at room temperature. Due to the different sedimentation rates of glomerulosa and fasciculata–reticularis cells (4.7 and >7 mm/h, respectively), more than 95% of the glomerulosa cells are found in the upper fractions, whereas fasciculata–reticularis cells can be collected close to the bottom of the gradient.

Since only a limited number of single cells (not more than $2.5 \times 10^7 - 5 \times 10^7$) can be separated by this method, the cell suspension needs to be washed and filtered thoroughly before unit gravity sedimentation. Therefore, the size-fractionation of adrenocortical cells is time consuming and limited by the small amount of cells yielded by each run. Furthermore, it has been shown that *in vitro*, zona glomerulosa cells rapidly lose their ability to secrete aldosterone and behave functionally and morphologically exactly like cultures from the two other cortical zones (Crivello *et al.*, 1982; Penhoat *et al.*, 1988).

Bovine adrenomedullary cells

The adrenal medulla has been widely used as a model for studies of functional properties of the sympathetic nervous system. The use of isolated bovine chromaffin cells in primary culture has greatly facilitated the investigation of the synthesis and secretion of catecholamines and opioid peptides, and the mechanisms of the stimulus-secretion coupling. Under *in vitro* conditions, the secretion of catecholamines from bovine chromaffin cells by exocytosis appears to be triggered by the influx of extracellular Ca^{2+} and the resultant rise in intracellular free Ca^{2+} levels. Depolarisation caused by treatment with high K^+ (55 mM) or by activation of nicotinic receptors on chromaffin cells leads to Ca^{2+} influx and the exocytotic release of catecholamines. Bovine medullary cells in primary monolayer cell culture can be distinguished morphologically from cortical and other non-chromaffin cells on the basis of their rounded shape and highly refractile halo, which is in contrast to the fibroblast-like shape of cortical cells with large nuclei and flattened, elongated bodies (Unsicker & Müller, 1981). In addition, staining with neutral red is specific for chromaffin cells (Wilson & Viveros, 1981; Unsicker & Müller, 1981), whereas contaminating fibroblasts can be identified by fibronectin staining (Danielsen, Larsen & Gammeltoft, 1990). Using electron microscopy, chromaffin cells can be identified by the presence of abundant catecholamine storage vesicles (Wilson & Viveros, 1981).

Isolation and culture of bovine chromaffin cells

Generally, for isolation and purification of bovine chromaffin cells, the protocol for preparation of bovine adrenocortical cells can be applied. However, due to the different structure of the bovine medulla, enzymatic digestion of small pieces of minced medullary results only in very low cell yields. The preparation of medullary cells should be carried out at room temperature, since it has been shown that medullary cells which are prepared in the cold show a very high release of catecholamines (Kilpatrick et al., 1981). In order to substantially increase the yield of adrenomedullary cells, an additional enzymatic perfusion step can be included as described by Wilson & Viveros (1981) and Wilson & Kirshner (1983).

Protocol

1 Excise the adrenal glands, including periadrenal fat, adrenal vessels, and part of the vena cava, from the surrounding tissue, making sure that the capsule remains intact.

2 Remove the periadrenal fatty tissue, and cannulate the adrenal vein with a 3.25 mm cannulation tube connected with a polyethylene tube to a 10 ml syringe filled with perfusion medium (145 mM NaCl, 5.4 mM KCl, 1.0 mM NaH_2PO_4, 11.2 mM glucose, 15 mM HEPES, 40 µg/ml of gentamycin (pH 7.4), 37 °C).

3 Apply a light pressure to the syringe in order to exclude points of leakage. Then, make multiple 1 mm-deep slits in the adrenal cortex.

4 Wash out the red blood cells by perfusing the adrenal gland using a perfusion pump for 10 min, with an increasing flow rate from 3 to 10 ml/min. This results in a low flow of the washing medium from the multiple slits in the adrenal gland, and at the end of this step, the perfusion medium should be clear of red blood cells.

5 For digestion, add collagenase II (1.0 mg/ml) and DNase I (15 μg/ml) to the perfusion medium, and perfuse the adrenal gland with 20 ml of recirculating solution for 10 to 15 min at a flow rate of 10 ml/min at 37 °C.

6 Change the digestion medium and perfuse the gland for another 10 to 15 min as described above.

7 After washing and perfusion with digestion medium, transfer the adrenal glands to a petri dish, and dissect longitudinally. The medullae can now be easily scratched from the remaining cortical tissue and collected.

8 Mince the chromaffin tissue finely, and incubate with 100 ml of PBS containing collagenase II (1.0 mg/ml) and DNase I (14 μg/ml) in a shaking water bath at 37 °C for 30 min.

9 Collect the cells by centrifugation at 240 g for 10 min, and incubate again with 100 ml fresh enzyme solution for another 30 min.

10 After digestion, filter the cells twice, through sieves of 140 μm and 100 μm mesh opening with PBS, pellet and resuspend in 50 ml of perfusion medium.

11 For Percoll purification, mix 6 ml of a salt solution (1.54 M NaCl, 56 mM KCl, 56 mM glucose, 50 mM HEPES, 2% BSA and 0.01% phenol red, (pH 7.4)) with 54 ml of Percoll stock solution, add to the cell suspension, and centrifuge for 20 min at 49 000 g in 15 ml centrifuge tubes. After Percoll centrifugation, three cell fractions can be identified: a band of cellular debris near the top, a broad band of chromaffin cells extending to the middle of the gradient, and erythrocytes at the bottom of the vials.

12 Resuspend the chromaffin cell fraction in tissue culture medium (50% DMEM and 50% F12 including 10 mM HEPES, 3 mM glucose, gentamicin (52 μg/ml) and amphotericin B (0.5 μg/ml)), and pellet the cells for 5 min at 400 g.

13 Wash the cells two more times in tissue culture medium, resuspend in 100 ml of serum-containing complete growth medium (including 10% FCS) and count.

14 Plate out the cells at a density of 0.5×10^6–1.5×10^6 cells/cm^2 in prewarmed (for 2 h at 37 °C) culture dishes to facilitate cell attachment.

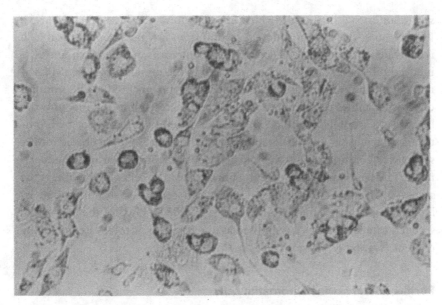

Fig. 5.3. Morphological appearance of bovine reticularis–fasciculata cells grown for 48 h in serum-containing medium (×200 magnification).

Establishment and maintenance of bovine medullary cell cultures

Cells are grown in complete growth medium (50% DMEM and 50% Ham's F12 including 5 mM HEPES, 28.3 mM $NaHCO_3$, 10 % FCS, 100 U/ml penicillin and 50 U/ml nystatin) at 37 °C in a moist atmosphere with 5% CO_2. Fibronectin-coated dishes can be used to improve attachment of the cells. For this purpose the culture dishes are preincubated with 10 μg of fibronectin in 0.5 ml adhesion medium (145 mM NaCl/5.4 mM KCl/1.0 mM NaH_2PO_4/11.2 mM glucose/15 mM HEPES/1.8 mM $CaCl_2$/0.8 mM $MgSO_4$, pH 7.4) for 1 h prior to plating (Wilson & Viveros, 1981). After 48 h the serum-containing medium can be replaced by serum free growth medium. Figure 5.3 shows the morphological appearance of bovine reticularis–fasciculata cells after 48 h in serum-containing medium. Primary cultures of bovine adrenal medullary cells can be maintained in the absence of serum for up to three weeks and still retain their ability to secrete catecholamines and other substances stored in the chromaffin vesicles (Kilpatrick et al., 1981). Under these conditions, increased cell density and decreased extent and frequency of medium replacement lead to improved maintenance of the cells (Wilson & Viveros, 1981). Catecholamine secretion can be stimulated by incubating the cells for 10 min at 37 °C in the presence or absence of various secretagogues in a high K^+-containing medium (Earle's balanced salt solution with 25 mM HEPES, 0.1 mM EDTA, 55 mM K$^+$). Furthermore, the catecholamine secretion from bovine adrenocortical cells is dependent on the extracellular Ca^{2+} concentration and the maximum K$^+$-induced catecholamine secretion is elicited in the presence of 2.5 mM Ca^{2+} in the stimulation medium (Dahmer, Hart & Perlmann, 1990). The amounts of catecholamines remaining in the cells and secreted into the medium can be assayed separately by HPLC with electrochemical detection or by monitoring the release of [^{3}H]norepinephrine from cells preincubated under serum-free conditions with this labelled catecholamine prior to the experiment. The secretion of [^{3}H]norepinephrine from the cells closely reflects the secretion of endogenous catecholamines and can be easily monitored by liquid scintillation counting (Kilpatrick *et al.*, 1981). Usually, the amount of secreted catecholamines is expressed as a percentage of total catecholamine content (Wilson & Viveros, 1981).

Purification of bovine chromaffin cells

The crude cell suspension obtained by the procedure described above usually contains only 50–75% chromaffin cells as determined by neutral red staining or electron microscopy. The contaminating non-chromaffin cells consist mainly of adrenocortical cells and a minor percentage of vascular endothelial, smooth muscle, and myelinated nerve fibre cells (Unsicker & Müller, 1981; Wilson & Viveros, 1981). A further purification of chromaffin cells can be achieved by density gradient centrifugation (Braley & Williams, 1980) or differential plating (Unsicker & Müller, 1981) of the cells. When crude medullary cell suspensions are subjected to density gradient centrifugation through 50% calf serum (Brooks, 1977) or a Percoll gradient (Goodfriend *et al.*, 1995), a purity of 85–95% can be obtained (Wilson & Viveros, 1981). Furthermore, the proliferation of non-chromaffin cells can be suppressed by addition of the antimitotic substances (e.g. 5-fluorodeoxyuridine (2.3 μg/ml) or cytosine arabinoside (2.8 μg/ml)) to the initial plating medium (Dahmer *et al.*, 1990). An even higher enrichment of medullary cells (97.5%) is achieved after five steps of differential plating, as described by Unsicker & Müller (1981). This method exploits the different adhesiveness of chromaffin and non-chromaffin cells to plastic and glass surfaces, and allows separation of adherent contaminating cells from the non-adhering medullary cells. Using this technique, the contamination of chromaffin cell cultures by cortical cells is very low (2.5%) but the loss of chromaffin cells during the procedure is immense, amounting to more than 95% of the total cells (Unsicker & Müller, 1981).

Rat adrenal cells

Since the rat adrenal gland contains a relatively large number of glomerulosa cells which can be easily separated from the two other cortical zones, rat adrenal glands are mainly used for the preparation of primary cultures of glomerulosa cells. Rat glomerulosa cells show a strong attachment to the thin capsule. Therefore, stripping of the capsule allows an efficient and easy separation of the glomerulosa from the flaccid two inner cortical zones. The following protocol describes the preparation of glomerulosa cells in primary culture (Yamaguchi et al., 1990). A similar protocol for the preparation of rat adrenal fasciculata-reticularis cells has been described by Mulay et al. (1995) and for the preparation of rat chromaffin cell cultures by Neely & Lingle (1992).

Source of rat adrenal glands

Usually, male or female Sprague–Dawley rats (200–250 g) are used for the preparation of rat adrenal cell cultures, but Wistar rats or Long–Evans rats can also be used. Immediately after the rats are killed, the adrenal glands are removed and kept in ice-cold PBS for further preparation. One rat adrenal gland weighs about 0.3 to 0.5 g and yields $1.0 \times 10^5 - 1.5 \times 10^5$ glomerulosa cells, $1.5 \times 10^5 - 2.0 \times 10^5$ fasciculata–reticularis cells, or 1.5×10^5 chromaffin cells. Usually 15–20 rats have to be sacrificed for one primary cell culture preparation.

Isolation of rat glomerulosa cells

Protocol

1 As soon as the adrenal glands are removed, clean away the fat under a dissecting microscope, then bisect and separate into capsular and decapsulated portions. For this purpose the adrenal capsule (with the attached glomerulosa cells) should be peeled off and cut into small fragments. Alternatively, another method of separating the capsular portion from the inner cortex and medulla is to incise the adrenal along one side and then to express the central zones by gentle, downward, rolling pressure with the thumb. These procedures allow clear macroscopic separation of white glomerulosa cells from the pink fasciculata–reticularis–medulla portion.

2 Incubate the capsular fragments in a shaking water bath for 1 h at 37 °C in 40 ml of M-199 containing collagenase I (1.5 mg/ml) and DNase I (50 μg/ml).

3 After enzymatic digestion, centrifuge the tissue fragments, and resuspend in ice-cold M-199, repeatedly pipetting up and down. Filter through a sieve of 50 μm mesh opening to remove larger fragments.

4 Wash the cell suspension several times, and resuspend in growth medium (M-199 with 0.2% BSA, 10% FCS, 100 μg/ml streptomycin and 50 U/ml nystatin).

5 Plate out the cells at a density of 1.5×10^4/ml in prewarmed culture dishes.

Establishment and maintenance of rat glomerulosa cell cultures

The protocol described above results in cultures containing 85–90% glomerulosa cells. For a further enrichment of glomerulosa cells (or fascic-ulata–reticularis cells when the decapsulated cortex is used), column separa-tion (McDougall *et al.*, 1979), unit gravity sedimentation (Braley & Williams, 1980) or Percoll purification steps (Chu & Hyatt, 1986) can be used. Usually, aldosterone and corticosterone release can be determined by radio-immunoassay of the conditioned medium after stimulation experiments. Rat glomerulosa cells in culture initially secrete aldosterone, but with increasing time, rising levels of corticosterone are produced (Crivello *et al.*, 1982).

Human adrenocortical cells

Human adult and fetal adrenocortical cells can be maintained in primary monolayer culture where they still retain many differentiated cell functions such as steroidogenesis and trophic responses to various growth factors and cytokines. In the human, the morphology and function of the fetal adrenal gland differs significantly from the adult adrenocortex. In the fetal adrenal gland the outer quarter of the cortex consists of the small definitive zone, which later develops into the adult adrenocortex and produces pre-dominantly cortisol. The major part of the fetal adrenal gland consists of the fetal zone, which shows a relative lack of 3β-hydroxysteroid dehydrogenase (3β-HSD) expression and therefore synthesises mainly dehy-droepiandrostenedione (DHEA) and its sulphate (DHEAS). Immediately after birth the fetal zone degenerates and the definitive zone steadily grows until it shows the typical zonation of the adult cortex during puberty. In the following section, a protocol for the purification and maintenance of adult human reticularis–fasciculata cells is given. However, the same protocol has also been used successfully for the preparation of fetal adrenocortical cells.

Source of adult human adrenal glands

Tissue from normal adult human adrenal glands can be obtained from patients who undergo total unilateral nephrectomy with ipsilateral adrenalectomy due to renal carcinoma. Immediately after surgical removal, the tissue should be dissected by the pathologist under sterile conditions and a sample of fresh non-necrotic adrenal tissue obtained. Alternatively, adult adrenal glands can be obtained *post-mortem* from patients who have given written consent for removal of organs for renal transplantation. The recommendations of the local ethical committee should be followed. The combined weight of the adrenal glands from human adults is about 8 g, but the weight and size of the glands varies considerably with age and physical condition. Usually, the amount of tissue that is available for tissue culture purposes is very small, and a careful tissue preparation must be performed in order to obtain enough cells for culture.

Isolation and culture of adult human adrenocortical cells

Protocol

1 Immediately after removal, place the adrenal tissue in a 50 ml centrifuge tube filled with 25 ml sterile ice-cold PBS containing gentamicin (52 μg/ml) and amphotericin B (0.5 μg/ml). The tissue container should be placed on ice in a second box and transferred to the laboratory as soon as possible.

2 Carefully remove the periadrenal fat from the adrenal tissue. Avoid crushing the adrenal tissue with large scissors or forceps.

3 Make a longitudinal incision of the gland, separate the brownish medulla from the cortex, and discard. The cortex, which represents 90% of the weight of the human adrenal gland, appears yellow due to the numerous large lipid droplets of adrenocortical cells. In contrast, the medulla is of a brown or dark red colour, due to the larger number of blood vessels . In the human adrenal gland, the boundary between cortex and medulla is not as clearly demarcated as in the bovine gland. However, human cortical tissue is of a more solid consistency than the tissue of the medulla, and the latter can be easily scraped off the yellow cortex with a scalpel.

4 In order to facilitate enzymatic cell separation, cut the tissue into very small fragments, and finely mince the remaining pieces with a scalpel.

5 For digestion, incubate the tissue in 40 ml of tissue culture medium (M-199, gentamicin (52 μg/ml), and amphotericin B (0.5 μg/ml)) contain-

ing freshly prepared collagenase II (1 mg/ml) and DNase I (200 μg/ml) for 45 min at 37 °C in a shaking water bath.

6 After digestion, filter the cells through a sieve of 100 μm mesh opening. For this purpose, pour the digested tissue onto the filter and gently stir with a sterile spoon while washing at 37 °C with 100 ml of culture medium containing antibiotics and 10% FCS. It is important not to use mechanical force, in order to avoid crushing of the cells with consequent cytotoxic effects of lysosomal enzymes.

7 Transfer the 100 ml cell suspension to two 50 ml Falcon tubes, and centrifuge for 5 min at 240 *g*. Resuspend the cells in 50 ml of tissue culture medium at 37 °C, and pellet again for 5 min at 240 *g* in order to separate the cell suspension from cell debris.

8 For Percoll purification from red blood cells and cell debris, resuspend in 10 ml of iso-osmotic Percoll solution (combine 10 ml of 1.5 M NaCl with 49.3 ml Percoll stock solution and add 40.7 ml double-distilled water at 37 °C), and centrifuge in a horizontal swing bucket centrifuge for 10 min at 730 *g*.

9 Wash the fasciculata–reticularis cells, which are concentrated in a layer on top of the Percoll solution, twice with tissue culture medium, in order to remove the remaining Percoll, and resuspend in a defined volume of growth medium. Using this Percoll centrifugation procedure, more than 70% of the red blood cells can be removed while cell viability still exceeds 90%.

10 Plate out the cells in serum-containing growth medium (M-199 supplemented with glutamine (290 ng/ml), gentamicin (52 μg/ml), amphotericin B (0.5 μg/ml), 10 % fetal calf serum and 5% horse serum) and incubate at 37 °C in a moist atmosphere with 5% CO_2.

Establishment and maintenance of adult human fasciculata–reticularis cell cultures

The yield of adrenocortical cells largely depends upon the quantity of adrenal tissue obtained, and varies considerably from preparation to preparation. Usually, the cells are plated in 24-well culture dishes (1 ml medium per well) at a density of approximately 4×10^4 cells/ml. After 24 h, 0.5 ml of complete medium is added to each well. At this time, the old medium should not be changed, in order to avoid aspiration of unattached cells. Thereafter, medium is replenished every day. Adult human adrenocortical cells exhibit a very good plating efficiency (more than 90% of the cells are attached after 48 h) and secrete larger amounts of steroids/cell than bovine cells. However, adult human fasciculata–reticularis cells do not grow as fast as fetal human or bovine adult adrenal cells, and confluence ($3 \times 10^4 - 4 \times 10^4$ cells/cm^2) is reached after approximately 5 days of culture in serum-containing medium. The primary monolayer adrenocortical cell cultures can be maintained in the same tissue culture dish for up to 120 h, while still retaining their differentiated cell function and morphology.

Troubleshooting

1 *After enzymatic tissue digestion, the cell suspension does not show a milky appearance and undigested tissue fragments are present.* Scrape off the tissue from the capsule in very small portions, and completely mince the removed fragments with a scalpel in order to make sure that no larger tissue fragments remain. Increase the collagenase concentration and/or digestion time.

2 *Only a small cell pellet is obtained after centrifugation of the filtered cell suspension (too much cell debris).* Reduce the digestion time and/or collagenase concentration. In order to avoid crushing the cells, do not use mechanical force to filter the digested tissue through sieves.

3 *Following centrifugation, the cell pellet obtained from the filtered cell suspension cannot be resuspended.* To avoid clumping of the cells, be sure to resuspend the cell pellet in a small amount of PBS after each washing step. Carefully increase the DNase concentration during the digestion process, in order to remove any free DNA fragments, which will enhance cell clumping.

4 *No clear separation of the cell layer is obtained after Percoll centrifugation.* Ensure that all PBS is removed before resuspending the cells in Percoll solution, since any PBS remaining will disturb the separation. Only use prewarmed (37 °C) Percoll solution.

5 *Cells cannot be easily aspirated after the Percoll centrifugation.* Increase the DNase concentration during the digestion process in order to avoid clumping of the cells. If large amounts of Percoll solution have been aspirated together with the cells, include additional washing steps in order to completely remove the remaining Percoll. Use only careful, intermittent suction when aspirating the cells.

6 *The cell viability before plating is below 75%.* Try to keep the preparation time as short as possible. Keep the tissue and cell suspension at 4 °C at all times (except during the enzymatic digestion). Avoid mechanical force during filtering of the cell suspension. Decrease the digestion time and/or enzyme concentration. Avoid Percoll contamination of the cells in order to reduce the cytotoxic effect of Percoll.

7 *Too much cell debris or red blood cells remain in the cell culture after the first wash.*

Decrease the digestion time and/or enzyme concentration in order to avoid cell fragmentation. Avoid mechanical force during filtering of the cell suspension. Add another washing step.

8 *After washing the cell culture for the first time, only a small number of cells remain attached.* Try other culture plates, since cells attach differently to tissue culture plasticware from different manufacturers.

9 *After 24 h, the cell density varies within each well.* Ensure that the cell suspension is continuously mixed while plating.

10 *After initial attachment, the cells die and only cell fragments remain attached to the plastic surface.* Try to keep the preparation time as short as possible. Keep the tissue and cell suspension at 4 °C at all times (except during enzymatic digestion). Avoid mechanical force during filtering of the cell suspension. Decrease the digestion time and/or enzyme concentration. Avoid Percoll contamination of the cell suspension after the Percoll centrifugation step.

11 *Cells don't grow and the viability drops after initial attachment of the cells, or culture medium is acidic or milky.* Check for contamination of the cell culture. Ensure that sterile conditions are maintained during all steps of the preparation.

References

Braley, L.M. & Williams, G.H. (1980). The effect of unit gravity sedimentation on adrenal steroidogenesis by isolated rat glomerulosa and fasciculata cells. *Endocrinology*, **106**, 50–5.

Brooks, J.C. (1977). The isolated bovine medullary chromaffin cell: a model of neuronal axcretion–secretion. *Endocrinology*, **101**, 1369–78.

Chu, F. & Hyatt, P.J. (1986). Purification of rat adrenal glomerulosa cells by Percoll density gradient centrifigation and the isolation of a population of cells highly responsive to adrenocorticotropin. *J. Endocrinol.*, **109**, 351–8.

Crivello, J.F., Hornsby, P.J. & Gill, G.N. (1982). Metyrapone and antioxidants are required to maintain aldosterone synthesis by cultured bovine adrenocortical cells. *Endocrinology*, **111**, 469–79.

Dahmer, M.K., Hart, P.M. & Perlmann, R.L. (1990). Studies on the effect of Insulin-like growth factor-I on catecholamine secretion from chromaffin cells. *J. Neurochem.*, **54**, 931–6.

Danielsen, A., Larsen, L. & Gammeltoft, S. (1990). Chromaffin cells express two types of insulin-like growth factor receptors. *Brain Res.*, **518**, 95–100.

Goodfriend, L.T., Ball, D.L., Elliott, M.E. & Shackleton, C. (1995). Lead increases aldosterone production by rat adrenal cells. *Hypertension*, **25**, 785–9.

Hornsby, P.J. & McAllister, J.M. (1981). Culturing steroidogenic cells. *Methods in Enzymology*, **206**, 371–80.

Kilpatrick, D.L., Ledbetter, F.H., Kirshner, F.H., Slepetis, R. & Kirshner, N. (1981). Stability of bovine medullary cells in culture. *J. Neurochem,* **35,** 679–85

McDougall, J.G., Williams, B.C., Hyatt, P.J., Bell, J.B.G., Tait, J.F. & Tait, S.A.S. (1979). Purification of dispersed rat adrenal cells by column filtration. *Proc. Roy. Soc. Lond. Series B: Biol. Sci.,* **206,** 15–31.

Mulay, S., Vaillancourt, P., Omer, S. & Varma, D.R. (1995). Hormonal modulation of atrial natriuretic factor receptors in adrenal fasciculata cells from female rats. *Can. J. Physiol. Pharmacol.,* **73,** 140–4.

Neely, A. & Lingle, C.J. (1992). Two components of calcium-activated potassium current in rat adrenal chromaffin cells. *J. Physiol.,* **453,** 97–131.

Penhoat, A., Jaillard, C., Crozat, A. & Saez, J.M. (1988). Regulation of angiotensin II receptors and steroidogenic responsiveness in cultured bovine fasciculata and glomerulosa adrenal cells. *Eur. J. Biochem.,* **172,** 247–54.

Simonian, M.H., White, M.L. & Gill, G.N. (1982). Growth and function of cultured bovine adrenocortical cells in a serum-free defined medium. *Endocrinology,* **111,** 919–27.

Unsicker, K. & Müller, T.H. (1981). Purification of bovine adrenal chromaffin cells by differential plating. *J. Neurosci. Methods,* **4,** 227–41.

Wilson, S.P. & Kirshner, N. (1983). Preparation and maintenance of adrenal medullary chromaffin cell cultures. *Methods Enzymol.,* **103,** 305–12.

Wilson, S.P. & Viveros, O.H. (1981). Primary culture of adrenal medullary chromaffin cells in a chemically defined medium. *Exp. Cell Res,* **133,** 159–69.

Yamaguchi, T., Naito, Z., Stoner, G.D., Franco-Saenz, R. & Murlow, P.J. (1990). Role of the adrenal renin–angiotensin system on adrenocorticotropic hormone- and potassium-stimulated aldosterone production by rat adrenal glomerulosa cells in monolayer culture. *Hypertension,* **16,** 635–41.

6

Leydig cells

Frank Chuzel, Hervé Lejeune and José M. Saez

Introduction

Long-term experiments with animals have contributed to our understanding of the hormonal control of testicular steroidogenesis and of the potential of various agents as modulators of testicular function. In addition, it has become clear that the Leydig cell steroidogenic response to luteinising hormone (LH) and human chorionic gonadotrophin (hCG) can be modulated by many hormones other than LH and hCG, indicating that the *in vivo* regulation of androgenic secretion by the testes is multihormonal and multifactorial (Saez, 1994). However, given the heterogeneity of the cellular population and the cell–cell interactions in the testis, it is not clear whether the increased or decreased steroidogenic capacity observed following the administration of hormones or other factors is due to a direct effect on the Leydig cell or whether this effect is mediated by another cell type.

Primary cell culture provides the opportunity to dissect and separate the various actions of substances influencing growth and differentiation. In the isolated cell system, the complicated *in vivo* situation can be broken down into elements, allowing an investigation into whether the effect of a given agent is due to a direct or indirect action on the cells. Primary culture thus plays a major role in studies of cell and organ physiology and is also a convenient tool for correlating morphologic with biochemical changes (e.g. in cell differentiation).

In order to study the direct effect of any substance on Leydig cell function, it would seem necessary to work with pure preparations of Leydig cells cultured in a chemically defined medium in which they retain their specific functions for extended periods of time. With the establishment of clonal cell lines, some laboratories have chosen to investigate the properties of these, rather than to work with primary cultures (Hoelscher & Ascoli, 1996). Although cell lines do resemble their normal counterparts in many aspects and can be easily propagated in culture, they do not retain all the differentiated functions of normal Leydig cells and, for example, the major steroids produced from endogenous precursors are progesterone and 20α-hydroxyprogesterone rather than testosterone.

Selection of model system and source of cells

Porcine Leydig cells

The anatomy of the testis, the volume occupied by the interstitial tissue, and the relative volume of Leydig cells within the interstitial tissue, vary from one species to another and, within the same species, with age. These variations will affect both the yield of interstitial cells from each testis and the percentage of Leydig cells within the interstitial cell preparation. In the pig, interstitial tissue occupies up to 80% of the whole testis volume during the first month of the postnatal life, and Leydig cells represent about 70% of this tissue. In contrast, in the rat, Leydig cells comprise only 3% of the testicular tissue (Mather, Saez & Haour, 1981). However, the structure of porcine testes is firmer than that of rat testes, and isolation of interstitial cells from the porcine testis requires more extensive enzymatic digestion (increased enzyme concentration and time of digestion) than does the isolation of interstitial cells from rat or mouse. Nevertheless, the porcine testis provides an increased cell yield per gram of tissue and greater purity of the cell preparation than the rat testis.

The second advantage of porcine Leydig cells compared to those of rat origin is their behaviour in culture. Rat Leydig cells cultured in a chemically defined medium lose most of their LH/hCG receptors, and some steroidogenic enzyme activities within the first three days in culture, whereas porcine Leydig cells cultured in the same medium retain most of their differentiated functions for at least one week. Thus, whereas rat Leydig cells, freshly isolated or in very short-term culture, can be useful to study the acute effect of factor addition, porcine Leydig cells are a more convenient model to investigate the long-term effects on cell proliferation and differentiation.

Porcine Leydig cells are most conveniently prepared from porcine testes obtained from a pig farm at the time of routine castration of piglets at 3–4 weeks of age, which corresponds to the maximum of the second wave of porcine Leydig cell development (Van Vorstenbosch, Colenbrander & Wensing, 1984). Such cells have high numbers of LH/hCG receptors (Peyrat, Meusy-Dessolle & Garnier, 1981), and secretion of testosterone is high.

Human Leydig cells

Since species-specific differences in the response of Leydig cells to several effectors have been reported (Saez, 1994), the potential effects of an agent on human Leydig cell function can be confimed only by testing it on human Leydig cell cultures.

The most critical issue is to obtain human testes for Leydig cell isolation. In order to isolate a sufficient amount of human Leydig cells, one or both whole testes are necessary, because the total number of Leydig cells in adult human testes is relatively low (5×10^8) (Nistal *et al.*, 1986) and the procedure of purification does not yield more than $2 \times 10^7 - 8 \times 10^{-7}$ Leydig cells.

Depending on local legislation, it may be possible to obtain human testes from adult men, following cerebral death, when kidneys and other organs are removed for transplantation. Such a procedure was possible in France until July 1994. The advantage of this source of tissues is that both testes are available and most of the subjects are young. A disadvantage is that the availability of the testes cannot be planned for in advance. Thus, the staff and the culture materials must always be available for performing the Leydig cell isolation procedure. The second source of human testes is therapeutic orchidectomy. However, the indication of the orchidectomy, age, hormonal treatment and illness represent limitations to the possibility of Leydig cell isolation. The most frequent indication of therapeutic orchidectomy is prostate carcinoma when both testes can be obtained. However, even if the patients do not receive hormonal treatment before castration, ageing (Neaves *et al.*, 1984) and illness induce testicular alterations of different intensity among such patients. Depending on the degree of fibrosis of the testis, this results in irregular action of the collagenase, and irregular Leydig cell yield. However, when Leydig cells have been isolated in sufficient numbers, *in vitro* testosterone production is similar to that obtained from Leydig cells obtained from subjects following recent cerebral death.

Chemicals and reagents

Tissue culture flasks (Becton-Dickinson, NJ, USA)

Tissue culture medium. Powdered Dulbecco's Modified Eagle's Medium (DMEM), Ham's F-12 medium (F-12), Ham's F10 $\times$10, phosphate buffered saline (PBS) and fetal calf serum (Gibco BRL, Life Technology, Paisley, Scotland)

Medium F : DMEM/F12 (1:1) containing 15 mM HEPES, pH 7.4, supplemented with 1.2 mg $NaHCO_3$, 16 µg gentamycin, 100 U penicillin, 100 µg streptomycin, and 50 U nystatin per ml of medium. Staining solution: nicotinamide-adenine dinucleotide (NAD) (0.47 mg/ml), dehydroepiandrosterone (0.1 mg/ml) and nitroblue tetrazolium (200 μg/ml) in medium F.

Penicillin-streptomycin, nystatin, trypsin-EDTA (BioMerrieux, Marcy l'Etoile, France)

Collagenase (Serva, Heidelberg, Germany)

Bovine serum albumin, DNase I, soybean trypsin inhibitor, collagen type I, fibronectin, laminin, porcine insulin, bovine transferrin, vitamin C and vitamin E (Sigma Chemical Co., St. Louis, MO, USA)

Percoll (Pharmacia Fine Chemicals, Uppsala, Sweden).

Human chorionic gonadotropin (hCG) (Sigma Chemical Co., St. Louis, MO, USA)

Fetal calf serum (Gibco BRL, Life Technology, Paisley, Scotland)

All chemicals should be of analytical grade.

Isolation and culture of porcine Leydig cells

This procedure is a mild enzymatic digestion that yields relatively pure preparations of interstitial cells. The conditions and time shown are chosen for use with immature (3 to 4 weeks of age) pig testes. Preparations contain the pooled testes of males from two to three litters. The reproducibility between preparations is normally excellent, with little variation seen in parameters other than total testis weight and cell yield per testis. Porcine testes should be obtained from a local pig farm at the time of routine castration, and transported to the laboratory in cold medium F. No more than 2 h should elapse between castration and the beginning of preparation of the cells. After collection, the testes should be washed several times with fresh medium before further processing, in order to avoid bacterial contamination. All of the subsequent procedures should be carried out under sterile conditions.

Protocol

1 Decapsulate the testes, remove the rete testis (Fig. 6.1), and finely slice the testicular tissue with a razor blade. Wash the slices two or three times with fresh medium, and then incubate under constant slow mechanical agitation at 33 °C for 90 min in medium F containing collagenase (0.75 mg/ml), soybean trypsin inhibitor (1 μg/ml) and DNase I (10 μg/ml) (a total of 400 ml for 15 testes). DNase I is added to prevent clumping of tissue due to 'stickiness' of released DNA. Trypsin inhibitor is included in the enzymatic digestion step to avoid tryptic digestion of cell surface

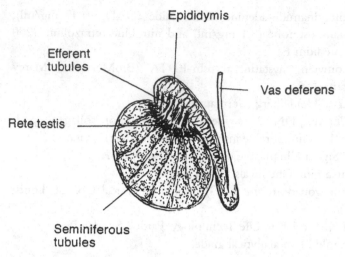

Fig. 6.1. Diagram of the testis, showing the anatomical location of the rete testis for removal prior to the isolation of Leydig cells.

receptors. The relatively low enzyme concentration and large volume ratios will minimise the incidence of cell damage during digestion.

2 Filter the dispersed cells through a nylon sieve (400 mesh) to remove non-dissociated pieces of tissue, and pellet by centrifugation at 200 *g* for 5 min at 4 °C. Resuspend the cell pellet in 50 ml of medium in a sterile conical centrifuge tube by pipetting gently. Then allow sedimentation for 5–10 min to eliminate the seminiferous tubules, which will settle by gravity. Retain the supernatant, and perform a second sedimentation for 30 min. Recover the crude suspension of cells, and separate the Leydig cells by centrifugation at 200 *g* for 5 min at 4 °C. Combine the final pellets, remove representative aliquots, and count cells in a Coulter counter. The cell density should then be adjusted to about 2.5×10^7 cells/ml.

3 Purify the Leydig cells using a discontinuous Percoll gradient. The starting solution should be 90% Percoll (Percoll/10 X Ham's F-10, 9:1 v/v). This solution is then diluted with medium F to achieve the following percentage of Percoll : 60%, 40%, 34%, 26% and 21% with a density of 1.082, 1.057, 1.048, 1.040 and 1.032 g/ml, respectively. These solutions should be prepared in the morning of the day of cell preparation and kept at 4 °C. The required volumes of these solutions (5, 7, 7, 5 and 8 ml, respectively), should be poured gently into a 50 ml conical tube. A suspension of 10^8 cells in 5 ml of medium F is then applied to the top of the gradient. After centrifugation at 1500 *g* for 30 min at 4 °C, several bands should be

Table 6.1. *Activity of isolated porcine Leydig cells before and after Percoll gradient purification*

		hCG-induced production		
		Testosterone $\left(\dfrac{\text{ng/2h/}}{10^6\text{cells}}\right)$	cAMP (%) $\left(\dfrac{\text{pmol/30min/}}{10^6\text{cells}}\right)$	3β-HSD cells (%)
Crude preparation		24±4	22±5	45±4
Percoll gradient bands				
	Density (g/ml)			
I	1.032–1.040	0.5±0.05	2±0.1	15±2
II	1.040–1.048	2±1	4±1	6±1
III	1.057–1.082	49±3	58±4	86±5
IV	>1.082	>0.1	>0.2	0

Note:
Values expressed as Mean±S.D.

observed, as illustrated in Table 6.1. The upper band at the top of gradient, between 21 and 26% Percoll, will be composed mainly of cellular debris. A second band between 26 and 34% Percoll, will contain about 30% of the total cell population, but most of these will not be Leydig cells, and the specific markers of these cells (binding of ^{125}I-hCG, 3β-HSD activity and production of testosterone) will therefore be negligible. A third main band, observed between the 40 and 60% Percoll layer ($d = 1.057/1.082$ g/ml) will contain about 40% of the total number of the cells, of which most (80–90%) should be Leydig cells, as defined by the specific markers described above. The last band, found at the bottom of the tube, is composed essentially of red blood cells.

4 To culture the Leydig cells, collect the cells of the third band ($d = 1.057/1.082$ g/ml), dilute with three volumes of medium F and centrifuge at 200 g for 10 min. Wash the pellet once with medium F, and finally resuspend in medium F supplemented with insulin (10 μg/ml), transferrin (10 μg/ml), vitamin C (0.1 μM), vitamin E (10μg/ml) and 0.2% fetal calf serum.

5 Seed the cells either into culture flasks (75 cm²) at a concentration of 1.5×10^7 cells/flask in 20 ml of medium, or into 6- or 24-multiwell culture dishes at 2×10^6 and 0.7×10^6 cells/well respectively. Maintain the cells in a controlled humidified atmosphere of 5% CO_2 in 95% air at

33 °C. After 48 h, remove the medium and replace it with fresh complete medium without serum, with or without the factor to be investigated.

Isolation and culture of human Leydig cells

Testes should be removed from adult men following recent cerebral death, at the time kidneys are removed for transplantation. They should be transported to the laboratory in cold medium F. Once decapsulated, a small fragment of testicular tissue should be cut off and placed in Bouin's solution for histological examination. This allows for checks that the testes have a normal histological pattern.

Protocol

1 Using forceps, gently dissect the tissue from the two testes, then digest for 90–120 min with constant agitation at 37 °C in 200 ml of medium F containing collagenase (1 mg/ml), DNase I (10 μg/ml), and soybean trypsin inhibitor (1 μg/ml). Filter the digested tissue through a large nylon gauze (400 mesh), and centrifuge the cells at 200 g for 5 min. Resuspend in 200 ml of fresh medium F, and separate the seminiferous tubule fragments and interstitial cells by gravity sedimentation for 15 min. The supernatant may be subjected to another gravity sedimentation for 30 min to let small tubular fragments settle. Recover the suspended interstitial cells by centrifugation at 200 g for 5 min, resuspend in fresh medium, and count.

2 Purify the Leydig cells using the discontinuous Percoll gradient as described above. Apply a suspension of 10^8 interstitial cells in a volume of 5 ml onto the discontinuous Percoll gradient. After centrifugation at 1500 g for 30 min, discard the heterogeneous cellular material remaining at the top of the gradient and between the 21 and 26% Percoll layers. Three cell bands should be seen; 'L1' between 26 and 34% Percoll layers, 'L2' between 34 and 40% Percoll layers, and 'L3' between 40 and 60% layers. Since L1 cells secrete very small amounts of testosterone, the L1 layer should be discarded. The L2 layer (1.048<buoyant cell density <1.057) and L3 (1.057 <cell density <1.082) which contains the Leydig cells should be collected, diluted with 3 volumes of medium F, centrifuged at 200 g for 10 min, and counted with a Coulter counter. Omitting the 40% Percoll layer results in pooling of the L2 and L3 bands, thus maximising the recovery of Leydig cells with an acceptable degree of purity (see below). Resuspend the Leydig cell pellet in culture medium F containing human insulin (10 μg/ml), bovine transferrin (10 μg/ml), vitamin C (0.1 mM) and

vitamin E (10 μg/ml), without fetal calf serum, count and dilute to a final concentration of 10^6 cells/ml.

Initiation of cultures

We have observed that plating of human Leydig cells on plastic dishes requires high concentrations (5–10%) of fetal calf serum. Under these conditions, plating is very efficient, but cells rapidly lose their responsiveness to hCG and, in addition, there is a rapid proliferation of contaminant fibroblasts. To overcome these pitfalls, we have used plastic dishes coated with collagen type-I, fibronectin and laminin (Lejeune *et al.*, 1993) or with extracellular membrane from bovine corneal endothelial cells (Cudicini *et al.*, 1997).

Collagen–fibronectin–laminin-coated wells may be prepared by dissolving rat tail collagen-I in acetic acid (50 mM), adding to the wells (10 μg/cm²) and evaporating to dryness under a sterile air flow. The collagen-coated wells should be washed 3 times with 10 mM PBS (pH 7.3), before adding fibronectin (1 μg/cm²) and laminin (1 μg/cm²) in 10 mM PBS (pH 7.3), and evaporating to dryness under a sterile air flow. Finally, the wells should be washed with PBS, followed by F-12/DME, before the cells are plated.

Extracellular basal membrane may be prepared from bovine corneal endothelial cells, as described by Gospodarowicz (1984).

Leydig cells should be plated in 12-well culture dishes at a density of 10^6 cells/well (1ml/well). Cultures should be maintained at 33 °C in a humidified atmosphere of 5% CO_2–95% air for 12 to 24 h. The medium should then be removed and replaced with a further, fresh portion of the same medium with or without the factors to be studied.

Cell viability and characterisation

Following isolation on the Percoll gradient, the viability of both human and porcine Leydig cells, as determined by trypan blue exclusion, should be in excess of 95%. Leydig cells may be identified by histochemical staining for 3β–HSD activity using a slight modification of the procedure described by Steinberger, Steinberger & Vilar (1966). This involves incubating the cells at 33 °C for 4 h in 780 μl of staining solution; the steroid substrate should be dissolved in propylene-glycol. After incubation, the cells should be washed with PBS and fixed for 1 h in glutaraldehyde (2.5%). Following extensive washing in PBS at the end of the fixation period, the percentage of 3β-HSD positive cells may be determined by counting a minimum of five fields of about 500 cells each using a light microscope. Between 80 and 90% of

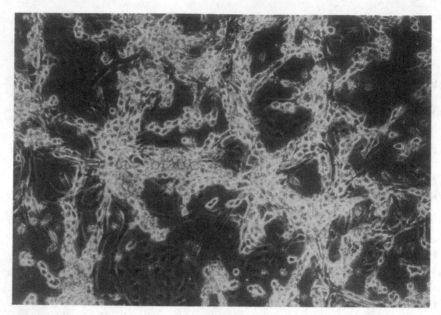

Fig. 6.2. Phase contrast photomicrograph of Percoll-purified porcine Leydig cells after 3 days in culture (×150 magnification).

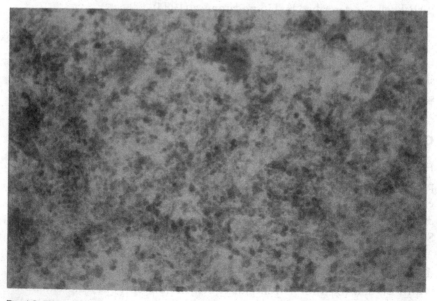

Fig. 6.3. Phase contrast photomicrograph of a mixed population of human Leydig and Sertoli cells after 96 h co-culture (×75 magnification).

Table 6.2. *Specific markers of porcine Leydig cells during culture*

Day	LH/hCG receptors		hCG-induced production	
	number	mRNA	Testosterone	cAMP
0	100	100	100	100
2	52±3	32±4	82±5	75±6
3	60±6	48±4	89±6	87±8
5	92±8	88±7	102±10	98±9

Note:
All values are expressed as % (mean±S.D.) of the value obtained for freshly isolated cells after Percoll purification (Day 0).

porcine isolated cells should stain positive. With human cells, the number of positive cells and the intensity of staining is higher in L3 ($\cong$80% positive cells) than in L2 cells ($\cong$60% positive cells), but human Leydig cells stain less intensely than those of porcine origin. This phenomenon has also been observed by others (Simpson, Wu & Sharpe, 1987) when comparing Percoll-purified human and rat Leydig cells.

Plating efficiency and characterisation of cultures

The plating efficiency for porcine Leydig cells after 48 h of culture varies between 60 and 80%, whereas the corresponding value for human Leydig cells after 24 h of culture is 45–65% on collagen–fibronectin–laminin coated dishes and 60–80% on basal membrane-coated dishes.

After the initial plating period, porcine Leydig cells are round and often associated in clusters (Fig. 6.2); about 90% are 3β-HSD positive. Those remaining are either Sertoli cells or peritubular cells, the proportion between these varying from one preparation to another (Fig. 6.3). When porcine cells are cultured in the chemically defined medium decribed above, cell proliferation as evaluated by ^{3}H-thymidine incorporation and cell counting is very low, but the cells retain some of their specific functions (i.e. LH/hCG receptor number and hCG responsiveness) (Chuzel *et al.*, 1995) (Table 6.2), although the mRNA levels of P450$_{scc}$ and P450c17 decline (Clark *et al.*, 1996).

After plating, human Leydig cells spread on the support, first exhibiting a stellate shape after 48 h of culture and then progressively changing to a fibroblast-like elongated shape (Fig. 6.4). Sertoli cells (Fig. 6.5), identified by

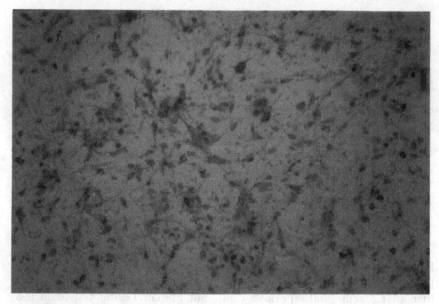

Fig. 6.4. Phase contrast photomicrograph of human Leydig cells after 96 h in culture, showing the adoption of a fibroblast-like morphology (×75 magnification).

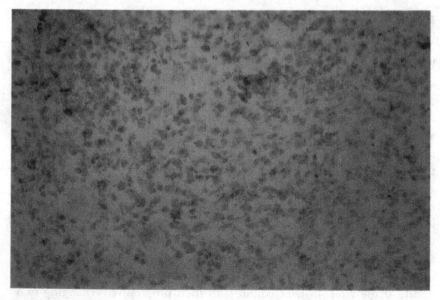

Fig. 6.5. Human Sertoli cells after 96 h in culture (phase contrast, ×75 magnification).

Table 6.3. *Long-term basal and hCG-induced testosterone secretion and acute cAMP and testosterone responsiveness of human Leydig cells to hCG*

Culture period	Long-term testosterone production		Acute responsiveness to hCG (10^{-9}M)	
	Basal (ng/48 h/10^6cells)	hCG (10^{-10}M) (% of basal★)	testosterone/2 h (% of basal★)	cAMP/30 min (% of basal★)
Human Leydig cells alone				
0–48 h	69±12	183±25	136±18	200±15
48–96 h	79±28	109±4	129±26	376±43
96–144 h	42±17	95±7	110±4	410±17
Human Leydig cells in coculture with human Sertoli cells				
0–48 h	113±9	139±24	131±3	121±8
48–96 h	193±17	99±5	105±10	354±33
96–144 h	366±98	106±5	107±7	362±25

Note:
★ : % (mean±S.D.) of the values for the same cells cultured without hCG for the same period of culture.

their typical granulations (Lipshultz, Murthy & Tindall,1982), account for less than 1% of the cells in mixed Leydig/Sertoli cell preparations. In contrast, these preparations contain between 4 and 8% of macrophages. These latter cells can be identified by immunostaining using a monoclonal anti-human monocyte CD14 antibody (Cudicini et al., 1997), according to the manufacturer's indications. The Leydig cell-enriched preparation can be depleted of macrophages by differential adhesion, by incubating for 30 min at 35 °C in a humidified atmosphere of 5% CO_2–95% air, in 75 cm^2 culture flasks (10 ml of cell suspension at 10^6 cells/ml for each flask). Non-adherent cells can then be removed and plated as a macrophage-depleted Leydig cell-enriched preparation, in which macrophages (CD14 positive cells) account for only 1–2% of the cell population.

When studied immediately after cell attachment, the basal testosterone secretion into the medium should be as high as 50–100 ng/10^6 cells/48 h. In contrast to this high basal level of testosterone secretion, hCG-induced testosterone secretion is of relatively low amplitude for human Leydig cells *in vitro*, amounting to about a 1.8 fold increase when the cells are cultured in the presence of 10^{-10}M hCG for 48 h (Table 6.3).

The most striking feature of adult human Leydig cells *in vitro* is their tendency to lose the ability to secrete testosterone unless they are co-cultured with Sertoli cells. The basal testosterone secretion then increases during the initial 96 h of culture (Lejeune *et al.*, 1993). When the co-culture is performed on a basal membrane, the secretion of testosterone is maintained for at least 1 month.

Cryopreservation

This procedure has been performed only with freshly isolated porcine cells. A suspension of 2×10^6 cells in 1 ml of complete culture medium containing 5% fetal calf serum should be cooled to 4 °C and then 0.1 ml of DMSO/ml of medium added. The cell suspension is placed in a cryo-freezer container (Nalgen, PolyLabo, Strasbourg, France) and kept overnight at −70 °C before being placed in liquid nitrogen. After defrosting, the plating efficiency should be about 60% and after 2 to 3 days in culture, the cells will show similar characteristics to cells cultured immediately after isolation.

Summary

Techniques have been described for the isolation and culture of human and porcine Leydig cells. The cells prepared by these procedures have been used to study the actions of many hormones and growth factors on specific cellular functions, including the production of testosterone in response to LH/hCG. They have enabled dissection of the mechanisms by which these factors modify Leydig cell functions. Finally, it has been demonstrated that a co-culture system of Leydig cells with other testicular cell types can be used to investigate cell–cell communication.

Troubleshooting

In addition to the specific details for the isolation and culture of each cell type described above, the following points should be taken into consideration.

1 *Contamination of porcine testes with bacteria or yeast* from the farm is a very frequent occurrence unless specific precautions are taken. Thus, careful washing of the whole testes after arrival at the laboratory is essential. If necessary the antibiotics in the culture medium should be adjusted to be maximally effective against specific contaminants.

2 *The activity of collagenase* will vary between different batches of the same material from different suppliers, and even from the same company. We

test several batches before purchasing the quantities required for 1 to 2 years' work. Assessment of collagenase batches may be based on parameters such as the yield of Leydig cells/gram of tissue, cell viability, binding of ^{125}I-hCG (receptor number/cell) and hCG-induced testosterone production.

3 *The use of chemically defined medium without serum* is necessary to investigate the specific functions of both porcine and human Leydig cells. Vitamin E supplementation has been shown to prolong the lifespan of porcine Leydig cells in culture. However, whether or not this vitamin plays the same role for human Leydig cells has not been evaluated. Vitamin C has been shown to be important as an antioxidant, for maintenance of the activity of several types of steroidogenic cells. Similarly, insulin, acting mainly through the type I IGF receptor, is important to maintain the response of LH/hCG receptors and the expression of genes for several steroidogenic enzymes (Chuzel *et al.*, 1996).

4 *Microscopic observation of the cells after the plating period* gives a good indication of the quality of the preparation. In addition, for porcine Leydig cells, measurement of ^{125}I-hCG binding is a rapid test to evaluate functionality.

References

Chuzel, F., Schteingart, H., Vigier, M., Avallet, O. & Saez, J.M. (1995). Transcriptional and post-transcriptional regulation of luteinising/ chorionic gonadotropin receptor by the agonist in Leydig cells. *Eur. J. Biochem.*, **229**, 316–25.

Chuzel, F., Clark, A.N., Avallet, O. & Saez, J.M. (1996). Transcriptional regulation of the lutropin/human choriogonadotropin receptor and three enzymes of steroidogenesis by growth factors in cultured pig Leydig cells. *Eur. J. Biochem.*, **239**, 8–16.

Clark, A.N., Chuzel, F., Sanchez, P. & Saez, J.M. (1996). Regulation of the messenger ribonucleic acid for P450 side chain cleavage, P45017α-hydroxylase/C17-20-lyase, and 3β-hydroxysteroid dehydrogenase in cultured pig Leydig cells. *Biol. Reproduc.*, **55**, 347–54.

Cudicini, C., Lejeune, H., Gomez, E., Bosmans, E., Ballet, F., Saez, J.M. & Jegou, B. (1997). Human Leydig cells and Sertoli cells are producers of interleukin-1 and interleukin-6. *J. Clin. Endocrinol. Metab.*, **82**, 1426–33.

Gospodarowicz, D. (1984). Preparation of extracellular matrices produced by cultured bovine corneal endothelial cells and PF-HR-9 endothelial cells: their use in cell culture. In *Methods for Preparation of Media, Supplements and Substrates for Serum-free Animal Cell Culture.* ed. D.W. Barnes, A. Sibasku & G.H. Sato, Vol.1. pp. 275–94. New York: Alan R. Liss.

Hoelscher, S.R. & Ascoli, M. (1996). Immortalised Leydig cell lines as models for studying Leydig cell physiology. In '*The Leydig Cells*', ed. A.H. Payne, M.P. Hardy & L.D. Russell. pp. 523–34. Vienna, Il USA: Cache River Press.

Lejeune, H., Skalli, M., Sanchez, P., Avallet, O. & Saez, J.M. (1993). Enhancement of testosterone secretion by normal adult human Leydig cells by co-culture with enriched preparations of normal adult human Sertoli cells. *Int. J. Androl.*, **16**, 37–54.

Lipshultz, L.I., Murthy, L. & Tindall, D.J. (1982). Characterisation of human Sertoli cells *in vitro*. *J. Clin. Endocrinol. Metab.*, **55**, 228–37.

Mather, J.P., Saez, J.M. & Haour F. (1981). Primary culture of Leydig cells from rat, mouse and pig: advantages of porcine cells for the study of gonadotropin regulation of Leydig cell function. *Steroids*, **381**, 35–44.

Neaves, W.B., Johnson, L., Porter, J.C., Parker, C.R. & Petty, C.S. (1984). Leydig cell numbers, daily sperm production, and serum gonadotropin levels in aging men. *J. Clin. Endocrinol. Metab.*, **59**, 756–63.

Nistal, M., Paniagua, R., Regadera, J., Santamaria, L. & Amat, P. (1986). A quantitative morphological study of human Leydig cells from birth to adulthood. *Cell Tiss. Res.*, **246**, 229–36.

Peyrat, J.P., Meusy-Dessolle, N. & Garnier, J. (1981). Changes in Leydig cells and luteinising hormone receptors in porcine testis during postnatal development. *Endocrinology*, **108**, 625–31.

Saez, J.M. (1994). Leydig cells: endocrine, paracrine and autocrine regulation. *Endocr. Revs.*, **15**, 574–624.

Simpson, B.J.B., Wu, F.C.W. & Sharpe, R.M. (1987). Isolation of human Leydig cells which are highly responsive to human chorionic gonadotropin. *J. Clin. Endocrinol. Metab.*, **65**, 415–22.

Steinberger, E., Steinberger, A. & Vilar, O. (1966). Cytochemical study of D5-3β-hydroxysteroid dehydrogenase in testicular cells grown *in vitro*. *Endocrinology*, **79**, 406–10.

Van Vorstenbosch, C.J.A.H.V., Colenbrander, B. & Wensing, C.J.G. (1984). Leydig cell development in the pig testis during the late fetal and early postnatal period: an electron microscopic study with attention to the influence of fetal decapitation. *Am. J. Anat.*, **169**, 121–35.

7

Thyroid follicular cells

Margaret C. Eggo

Introduction

The primary interest of most readers of this chapter is likely to be the examination of techniques to isolate functional cultures of thyroid follicular cells which can be used to examine the effects of environmental agents on thyroid function. Dietary goitrogens include agents such as fungicides (e.g. ethylene-*bis*-dithiocarbamates), antibiotics (e.g. sulphamethazine) and radio-contrast agents used diagnostically (Yamada *et al.*, 1975, Hill *et al.*, 1989). Functional cultures of thyroid cells can be used for screening purposes, and dose–responses and mechanisms of action of these agents can be determined (Drucker *et al.*, 1984; Gupta *et al.*, 1992, Divi & Doerge, 1994). Studies of thyroid growth have been used in a clinical setting to determine thyroid-stimulating antibodies and subtypes of these which can modify thyroid growth and/or function (Brown, 1995). The thyroid is an excellent research model for studying intracellular signalling mechanisms and interactions between signalling pathways (Maenhaut *et al.*, 1990). Thyroid function is stimulated by agents elevating cyclic AMP and inhibited by many growth factors, possibly by activation of certain protein kinase C isozymes (Eggo, 1993). Pathways involving *ras* activation (Al-alawi *et al.*, 1995) and those involving tyrosine phosphorylations are also being studied. Thyroid function is further regulated by iodide which, in excess, is a potent inhibitor of thyroid hormone secretion (Nagataki, 1975).

This chapter will concentrate on the culture of human thyroid cells, rather than on those from other species, although the cell isolation processes are similar. Primary culture of abattoir material does have advantages over culture of human thyroid cells because a large number of normal glands obtained from animals of the same sex and age can be pooled, thus limiting the heterogeneity that one sees with human surgical samples. The dis-advantage, however, is that there appear to be significant species differences in thyroid cell function and control. Many features of the thyroid gland have evolved in order to conserve iodide. Dietary availability of iodine thus governs iodide trapping efficiency and deiodinase levels and these parameters therefore vary between species. Herbivores differ from omnivores and carnivores in their daily iodide intake and each has evolved strategies to conserve dietary iodide. Rodent thyroids are too small for culture use and again the suitability of the model is questionable. Rats do not possess thyroxine-binding globulins, using instead the much weaker affinity of serum albumin to bind to thyroid hormones. This represents a fundamental difference in thyroid hormone delivery (the half-life of thyroxine (T4) in the rat is about 12 h compared to 7 days in humans) and is coupled with a much higher iodide trapping ability in the rat thyroid compared to the human (Davies, 1993). Another confounding problem with the rat model is the relative ease with which thyroid tumours are produced following goitrogen ingestion, and after a year of such treatment there is a 100% incidence. As far as can be determined from epidemiological data, this does not occur as readily in humans. Although unselected autopsy studies indicate that 50% of the population harbour thyroid nodules and up to 28% micropapillary carcinomas, this is not comparable to the levels found in rodents (Wheeler, 1994). It is therefore pragmatic to study human thyroid tissue rather than introduce animal models, the regulatory aspects of which may differ from those in the human model.

Characterisation and morphology

The thyroid gland is composed of at least four cell types, and the effects of the paracrine secretions of each of these on their neighbours are largely unknown. Moreover, the microenvironment will influence the way each cell type responds to hormones and other stimuli. The thyroid follicular cell cannot therefore be considered in isolation, and for this reason the characteristics and isolation of some of the other cells of the thyroid will also be briefly considered. An understanding of the individual cells within the thyroid gland, and an appreciation of their effects upon each other, have important parallels in any organ composed of several different cell types.

Thyroid follicular cells

The principal component (>70% of cells) of thyroid tissue is the thyroid follicular cell which is responsible for the synthesis of thyroid hormones. These cuboidal epithelial cells line closed spheroids known as follicles. The average diameter of a follicle is 200 μm, but this varies with the degree of stimulation that the gland is experiencing. The principal stimulator of thyroid growth and function is thyrotrophin, also known as thyroid-stimulating hormone (TSH). In the hyperactive gland the follicles are small and the follicular cells lining them are columnar, with dense cytoplasmic inclusions. In contrast, in the underactive state the follicles are much larger and the follicular cells lining them are flattened and appear squamous. At any time in the normal thyroid there is considerable variability in follicular size and functional activity. This is thought to be due to heterogeneity in the response of individual cells to TSH (Studer & Derwahl, 1995). The principal component of the follicle is thyroglobulin, a 660 kDa glycoprotein which contains within its matrix the thyroid hormones T4 and tri-iodothyronine (T3). Iodination of tyrosine residues and the coupling of two iodotyrosines to form T3 and T4 are mediated by the enzyme thyroperoxidase. Thyroglobulin, thyroperoxidase and the TSH receptor are three thyroid-specific proteins that identify the thyroid follicular cell.

The thyroid follicular cell is highly polarised in structure, and several studies have shown polarised secretion of proteins from the cells. The basal side of the cell is served by thyroid capillaries, and it is here that the recently cloned iodide transporter (Dai, Levy & Carrasco, 1996), the TSH receptor, and receptors for growth factors and insulin are located. The apical side, which faces the colloid, expresses thyroperoxidase and is also the site of uptake of thyroglobulin. There is an abundance of mannose-6-phosphate/type 2 IGF receptors on this membrane. To recreate functional (i.e. thyroid hormone-secreting) thyroid follicular units in culture, several criteria have to be fulfilled:

(i) *Structural reorganisation into three-dimensional follicles* which contain thyroglobulin, which involves maintenance of appropriate cell polarity. In fact, while this property is attractive to cell biologists, it is not essential for thyroid function as measured by assays of iodide metabolism and T3 and T4 secretion.

(ii) *The ability to trap iodide.* The iodide transporter should be able to maintain a concentration gradient, but in culture this property is frequently lost or not examined. The confirmation that iodide trapping is maintained in culture demonstrates that the synthesis and function of this labile protein has been preserved, and promotes confidence in the responses mediated through cell surface receptors.

(iii) *The ability to synthesise thyroglobulin.* Analysis of polysome profiles shows that 50% of protein synthesis in the thyroid is devoted to thyroglobulin production, although this high level is not usually maintained *in vitro*. Similarly, thyroperoxidase and TSH receptor synthesis should be maintained in culture.

(iv) *The ability to synthesise thyroid hormones.* Thyroid hormones are stored within the thyroglobulin molecule. When thyroid follicular cells are put into culture, these stores take considerable time to deplete (7–10 days in TSH-stimulated cells). The ability to synthesise thyroid hormones *de novo* should only be assessed when iodide dependence for thyroid hormone synthesis has been established, otherwise the effect examined is not synthesis but release.

Endothelial cells

The thyroid gland has a rich blood supply so endothelial cells of the vasculature are present in relatively high numbers (approximately 20% of cells in the thyroid). Endothelial cells are thought to be tissue specific and the characterisation of thyroid endothelium may be a worthy project in its own right. The relative contribution of the endothelium to thyroid mass can vary because in Graves' disease where autoantibodies are thought to activate the TSH receptor, a highly vascular gland is found. This is not simply due to the effects of hyperthyroidism, and in the rat thyroid, an angiogenic response to goitrogens precedes the follicular cell response. Whether TSH has any direct effect on endothelial cells, or whether their growth is mediated by paracrine growth factors produced by follicular cells, is unknown. Endothelial cells produce many autocrine and paracrine growth factors as well as thrombospondin, which promotes differentiation into capillaries but inhibits endothelial cell growth. Endothelial cells can be detected and isolated by their expression of Factor VIII.

C-cells

The C-cells are a minor component (~0.1% of cells within the thyroid gland) although the exact fraction can vary widely. They produce several bioactive molecules whose paracrine influences on their neighbours are unknown. In the human, C-cells are located within the basement membrane of the follicles but they are separated from the colloid by the follicular cells. They may also appear in the stroma between the follicles. They are slightly larger than follicular cells and although they are usually located randomly, they are found particularly in the middle third of the thyroid lobes. They secrete calcitonin, and a subset of C-cells (at least in rodents) produces the gut hormone somatostatin (Thomas *et al.*, 1994). Calcitonin secretion is stimulated by cyclic AMP which also regulates thyroid follicular cell function. Since cyclic AMP is not sequestered in cells but readily diffuses, crosstalk between thyroid follicular cells and C-cells may occur at the level of this second messenger.

Fibroblasts

Fibroblasts are the supporting cells within the thyroid gland, and are estimated to comprise 10% of the thyroid tissue mass. They produce fibrocollagenous tissue, mainly composed of collagen fibres, which supports nerves and blood vessels and separates follicles from each other by delicate septa. The problem for the cell culturist is that fibroblasts are very robust cells with a greater growth potential and tolerance of *in vitro* conditions than thyroid epithelial cells. They will overgrow the latter and precautions must be taken to eliminate them from follicular cell preparations. Like endothelial cells, fibroblasts may be tissue specific, and it is possible that thyroid fibroblasts detect and respond to different signals to fibroblasts from other sources. The role of the fibroblast in modifying its environment is only beginning to be appreciated; they are known to secrete insulin-like growth factors which, in the thyroid gland, are obligatory for differentiated function and growth. They also secrete insulin-like growth factor binding proteins, the composition of which can be sensitively regulated. Accordingly, the secretions of thyroid fibroblasts may influence the response of the thyroid follicular cells to hormones and growth factors.

Sources of cells

For primary cultures of thyroids from domesticated animals, the local abattoir is likely to be intrigued by a request for a small part of the carcass. Generally, the staff are very helpful although one should always be mindful that they could easily refuse. For human tissue, most thyroid surgeons are very helpful. The supply of normal thyroid tissue is, however, limited and experience suggests that the contralateral lobe of a diseased thyroid is frequently not as 'normal' as might be expected. Having established a research relationship with the local surgeon, it may be possible to obtain tissue from laryngectomies where the thyroid is presumably completely normal. Other sources of normal human thyroid tissue could be autopsy material, or that obtained by organ retrieval surgeons. Local ethical committee approval may be necessary for this retrieval. The thyroid gland is not especially fragile in this regard, and material that has been unprocessed for up to 48 h will usually yield healthy cell cultures. This practice is not to be encouraged, however, and in general collection of the tissue in a balanced salt solution at 4 °C and rapid processing are desirable. Multinodular thyroid goitres can sometimes achieve astonishing sizes (e.g. 500 g) and are frequently functional. Obviously, these are not truly normal, and their suitability for use depends on the question that is being asked in the proposed study. Cells derived from such tissue can be used, for example, in screening studies for goitrogens and in studies of iodide sensitivity, although studies on cell signalling pathways may be a little more suspect.

Calcium oxalate crystals are found in human thyroids and the amount is age dependent (Reid, Chang-Hyun & Oldroyd, 1987). They are insoluble and are removed with media changes. Such inclusions appear not to have any adverse effects on the isolated cells. They appear as large clear bodies in the initial cell isolate and are at least as large as thyroid follicles.

A small representative fragment of the tissue obtained from surgery may be fixed in 10% formol saline (3.7% formaldehyde, 0.9% saline). Following wax–embedding, this tissue can be sectioned and used for immunostaining or for an examination of ultrastructure and histology.

Chemicals, reagents and supplies

Sterile culture plates and plasticware: any reputable supplier (e.g. Beckton Dickinson; Nunc).
Collagenase: see text for details.
Culture medium: Ham's F12 (Coon's modification): custom-prepared.
Balanced salt solutions: these can be prepared in your own laboratory or purchased from any reputable supplier.
Newborn calf serum: any reputable supplier.

Processing of whole thyroid tissue

Figure 7.1 illustrates the location of the thyroid gland in relation to the other structures in the neck. In the case of human tissue, it is unlikely that the researcher would ever be faced with having to locate the thyroid gland, since surgeons will be removing the tissue. For tissue of animal origin, the thyroid gland is easily located in the neck by reference to the laryngeal prominence. The bilobed thyroid gland is an encapsulated discrete organ located caudally to this, extending from the second to the fourth tracheal rings.

Thyroid tissue should be transported to the laboratory in cold sterile

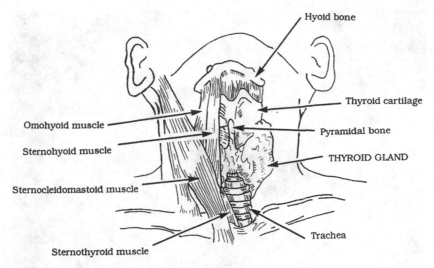

Fig. 7.1. The location of the thyroid gland in the neck, in relation to other structures. (Reproduced with permission from Greenspan & Strewler, 1997.)

Hanks' balanced salt solution (HBSS). Because thyroid tissue can be virally infected, a Hepa-filtered Class II containment cabinet should be used for all work with the tissue from this point onwards. The local arrangements for safe use of human tissue and its disposal should be followed. All subsequent manipulation of the tissue should be performed using sterile instruments and media.

Protocol

1 Trim the tissue free of fat and connective tissue on a sterile glass or plastic petri dish (Fig. 7.2). If processing a multinodular goitre, the nodules can be isolated intact and cultured individually. Connective tissue is white and fibrous and encapsulates the entire gland. The more thorough the trimming at this stage, the greater is the likelihood of a fibroblast-free culture. For tissue processing, a slicing action using opposed scalpel blades gives the best yield of undamaged cells. Other methods such as scissors, kitchen gadgets for slicing and electrical tissue choppers tend to tear the tissue rather than cut it, giving disappointing results. Cut the tissue until each piece is a cube of about 2 mm³.

2 Wash the chopped tissue several times in 40 ml ice-cold physiological salt solution; any such solution, as available, will keep the cells viable at this stage.

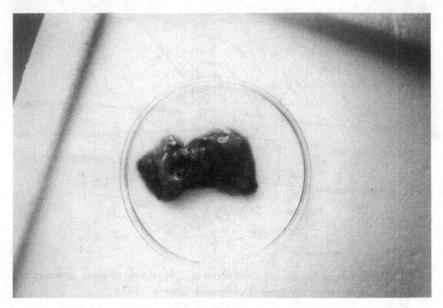

Fig. 7.2. Lobe of human thyroid gland (multinodular goitre), prior to trimming away connective tissue and processing (×0.5 magnification).

3 Allow the fragments to settle by gravity or by a 2 min centrifugation at 50 g , then aspirate and discard the supernatant containing blood cells, fat and connective tissue.

Digestion with enzyme

The chopped, washed thyroid tissue is incubated with 0.2% collagenase in HBSS (1:4 v/v). Sterile collagenase solution can be stored at −20 °C. The material is often difficult to filter due to high levels of particulate matter and a prefiltration through paper filters is advised to prevent blocking of the 0.2 μM sterilisation filters. Collagenases are metalloproteinases requiring Zn^{2+} and Ca^{2+} for their activity. The type and sometimes the batch of collagenase used is critically important. Some manufacturers, e.g. Boehringer Mannheim, Worthington (Freehold, NJ, USA) and the Sigma Chemical Co. will send sample aliquots of their available collagenase preparations, provided you agree to purchase a substantial quantity. They will reserve stocks of these trial batches until you have determined which gives the best yield of functional thyroid cells. Although this is a commendable practice, commitment to purchasing a large quantity of collagenase is a possible disadvantage. The advantage, however, is the elimination of a major source of variability. The reason for such batch-to-batch variability is that collagenase is not a pure preparation but a crude mixture of proteases from the extracellular filtrate of *Clostridium histolyticum*. The non-collagenase proteases are essential but ill-defined. A decision as to which type of collagenase to use initially is often confusing, as manufacturers have differing classifications. For Worthington, the type 2 collagenase is recommended, for Boehringer Mannheim the type A, and for Sigma the type 1 collagenase. Sigma now supplies four standardised collagenases (Sigma blend) which are said to provide a standardised product with no lot-to-lot variability. The four types vary in the ratio of collagenase : neutral protease activity.

Some workers use a mixture of Dispase™ (Boehringer Mannheim), a non-specific protease, with collagenase, but this does introduce another reagent variable. Early studies of thyroid cells in primary culture used trypsin to disaggregate the tissue, and viable cultures were obtained. However, this is a harsh treatment and the function of the thyroid cells is compromised. TSH receptors will regenerate but whether thyroperoxidase and the iodide transporter are equally robust is not known.

Isolation of follicles from thyroid tissue

For digestion of the thyroid tissue fragments, a fluted trypsinising flask (Bellco) can be used with a slowly turning 2 cm magnetic stirring bar. It is important that this agitation is gentle, as a whisking motion will rapidly damage the cells. Check also that the magnetic stirrer does not generate heat during its operation. This can create problems, particularly when incubation is performed in a warm room at 37 °C. Alternatively, end-over-end mixing in a large centrifuge tube on a rotary mixer gives some physical agitation but is not damaging. The digestion can be performed at 37 °C for 3 h, overnight at room temperature, or in the cold room. The latter digestion is usually incomplete and will require a further incubation at a higher temperature to effect good digestion. Providing the tissue is adequately chopped, a homogeneous digest should be isolated. There will be some clumps of connective tissue present, which are readily identifiable under the microscope as fibrous strand-like masses.

Protocol

1 After incubation, mix the homogenate with sterile HBSS, shake hard and pass through a stainless steel filter (250 μM) to remove undigested lumps of

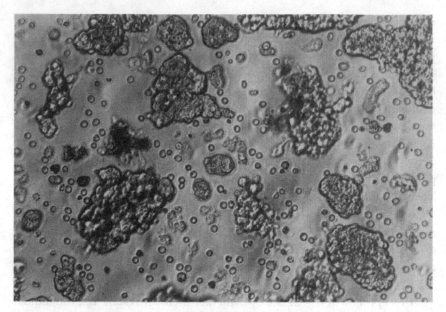

Fig. 7.3. Freshly-isolated human thyroid follicles. The small single cells are erythrocytes (magnification ×200).

tissue and aggregates of connective tissue which will be seen as white fibrous material on the filter. Sterile gauze may be used if a filter is not available.

2 Wash the homogenate through with at least 100 ml of HBSS and transfer to 50 ml centrifuge tubes. Pellet the follicles and cells at a low speed (300 *g* ; 1000 rpm, rotor radius 27 cm) in a swingout rotor for 10 min at 4 °C. Suspend the cells in a small amount of HBSS and transfer to 15 ml conical centrifuge tubes. Fill the tubes with HBSS and shake to resuspend the cell pellet thoroughly. Centrifuge the tubes for 2 min at 300 *g*. The supernatant contains some red blood cells, single cells such as fibroblasts, endothelial cells and cell debris.

3 Repeat the washing and centrifugation at least three times, at which point the thyroid follicles will appear as a buff/brown pellet overlayered with blood cells. If, at this stage, the distinct layer of follicles is not seen, repeat the washing step at least two more times. Figure 7.3 shows the appearance of thyroid follicles immediately following isolation.

Culture media

Various media have been used to culture thyroid cells. Initially, minimal essential media were supplemented with either fetal or newborn calf sera.

Cells plated in these media undoubtedly grow in culture, but their thyroid function is not maintained for long. To eliminate the need for serum, enriched culture media were developed in the early 1980s which, when supplemented with hormones specific for the cell type, allowed the culture of cells in chemically defined conditions. This has clear benefits for thyroid follicular cells. Firstly, the products secreted by the cells can be examined, whereas in serum-containing medium the enormous protein load of the serum precludes this type of study unless radioactive precursor molecules are added. Serum-free culture has allowed the isolation of many autocrine growth factors from thyroid follicular cells. Secondly, it should be recognised that thyroid follicular cells in the intact thyroid gland do not come into contact with serum, but are bathed in interstitial fluid, which contains only those substances in the blood that can pass through, or between, the endothelial cells. Lipid-soluble substances diffuse across the capillary walls with greater ease than lipid-insoluble substances probably by passing directly through the endothelial cells. In contrast, the large binding protein: hormone/ growth factor complexes within serum do not diffuse readily across the capillaries but free hormones and growth factors will cross. There are therefore many differences in the composition of the two liquids and serum is certainly not innocuous or even physiologic, indeed the presence of serum has been shown to inhibit thyroid function in a dose-dependent manner (Eggo et al., 1984; Roger et al., 1989).

In 1980 Ambesi-Impiombato, Parks and Coon published details of a new medium, based on Ham's F12M nutrient mixture, for the culture of the FRTL rat thyroid cell line under low serum (0.5%) conditions. The differences between this medium and the original F12 formulation are that (i) the concentrations of the amino acids are doubled and cystine has been substituted for cysteine, and (ii) large amounts of the antioxidant ascorbic acid are included. This may protect thyroid follicular cells from the oxidising environment produced during thyroid hormone synthesis. This medium has subsequently been used to culture the FRTL-5 cell line and has now been adopted by most thyroid research laboratories performing cell culture. A version of this medium (F12 Coon's modification) is available from the Sigma Chemical Co., which has a higher Zn^{2+} content than the original formulation. Whether this has any adverse effect has not been reported. The medium may be custom prepared in powdered form in aliquots, sufficient to make 1 litre of medium.

Because thyroid follicular cells are to be cultured under serum-free conditions, the quality of the water used to make up the medium must be extremely high (e.g. MilliQ™ water). Similarly, the glassware used to store the medium and the 0.2 μM filters must not contain any traces of detergent. Medium should be made up when required and not stored for long periods (i.e >3 weeks). Neither the ascorbic acid nor the glutamine in the media are stable even when refrigerated. The medium should not be warmed before use, and should not be exposed to unnecessary light.

Despite the inherent dangers of serum exposure, serum contains several necessary 'attachment factors' which are obligatory for thyroid cells to attach and spread on culture plates. Serum should therefore be included in the medium at the initial isolation stage (1% v/v), but removed at the first medium change. Newborn calf serum is used to supply the necessary adhesion factors while the cells attach and spread.

Hormones and other additives

The culture of thyroid follicular cells in chemically defined medium benefits from the addition of insulin, which is usually added at a sufficiently high concentration (10 μg/ml) to activate both the insulin and the IGF-I receptors. While this level is probably not harmful, lower concentrations (e.g. 1 μg/ml) are also effective. TSH is used at 100–300 μU/ml, which is optimal for maintenance of iodide uptake (Eggo et al., 1996). Recombinant TSH is now available and should be used where possible to obviate the criticism that the bovine preparation is contaminated with other pituitary growth factors and hormones. Ambesi-Impiombato et al. (1980) included a supplement of cortisol, somatostatin, glycyl-histidyl-lysyl acetate (rat liver growth factor) and transferrin for the growth of FRTL cells. The effect of these compounds on human thyroid cell function and growth is not yet established, although recent work using cortisol with porcine thyroid cells has shown stimulatory effects on thyroid iodide uptake. Transferrin and selenium are also frequently added to serum-free cultures. Transferrin transports iron into cells, while selenium is a component of several important enzymes. Since the procedures described in this chapter are for short-term (<1 month) cultures, addition of these factors to the cultures is unlikely to be necessary. However, if long-term cultures are desired, such additives may be important. A supplement of insulin, transferrin and selenium (ITS) can be purchased ready prepared from several companies. However, before complicating the media unnecessarily, it would be prudent to determine whether there is a real requirement for the additives.

Suppliers
Insulin: Sigma Chemical Co.
Thyrotrophin (TSH): Sigma Chemical Co.

Antibiotics

Penicillin (100 U/ml) and streptomycin (100 μg/ml) are added routinely to the medium. Gentamycin sulphate (50 μg/ml) can be substituted if bacterial contamination remains. It is reported to be active against mycoplasma, although for short term cultures this is not usually a consideration. Nystatin is helpful against yeasts of the candida variety and should be added to the plating medium at 50 μg/ml. This material is in suspension and media containing it should not be filtered. All antibiotics can be omitted from the medium at the first medium change. If mould develops on the cultures, it is impossible to eradicate and the cultures should be discarded.

Plating density

The optimal plating density for thyroid follicular cells is determined to some extent by the experiment to be performed. For growth studies, a suitable plating density is approximately 2×10^4 cells/cm^2. After centrifugation at 800 g for 2 min, 0.1 ml of the packed follicle pellet is suspended in 125 ml medium. This plating density will give sufficient space on the surface area of the culture plate for at least three cell divisions. For functional assessment studies, cells should be plated at 2.5 × this density. It is unnecessary to lyse erythrocytes at plating because they will be eliminated with time and the treatment necessary to remove them could be detrimental to thyroid cell function.

As an illustration of the effect of plating density on function, Fig. 7.4 shows the iodide uptake per μg protein for follicular cells plated at varying densities. When plated at 5×10^4 cells/cm^2, iodide uptake is high, and is retained at a four-fold dilution. However, it is dramatically reduced when plating density is reduced by a further four-fold. This fall is not reflected in the amount of protein on the plates (34 ± 4 μg in the 16-fold diluted plates compared to 59 ± 9μg in the four-fold diluted plates). The profile shown in Fig. 7.4 was obtained from cells 14 days after plating, but is also apparent at 4, 7 and 11 days. There is no significant increase in protein after the second medium change (7 days) at any of the plating densities, at which point cell growth will have ceased.

Culture plates

Tissue-culture plastic is treated so that it bears a positive charge which enhances cell attachment. Thyroid follicular cells are relatively adherent, and

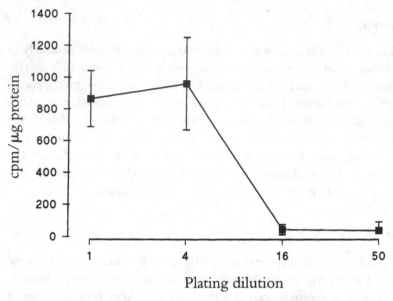

Fig. 7.4. Effect of plating density on the ability of thyroid follicular cells to take up iodide. Iodide uptake is corrected for μg protein from 14 day old thyroid cells plated at varying densities. Mean±SD, $n=6$. $1 = 5 \times 10^4$ cells/cm^2; $4 = 1.25 \times 10^4$ cells/cm^2; $16 = 3 \times 10^3$ cells/cm^2; $50 = 10^3$ cells/cm^2.

any brand of culture-ware should be suitable. Although collagen-coated plates promote cell attachment, there has not been any comparative study that conclusively proves better functional performance after such treatment. As such plates are time consuming to prepare and expensive to purchase, they will not be considered further. Matrigel™ (Beckton Dickinson), a soluble basement extract from tumour cells that gels to form a basement membrane of laminin, collagen IV, entactin and heparan sulphate proteoglycan, is purported to allow three-dimensional follicular structures to form and thus improve function. However, a thyroid follicle can also be formed on a plastic, tissue culture-treated dish as well as in an agarose gel, so Matrigel™ is not essential for such reorganisation (Frohlich, Wahl & Reutter, 1995). For morphological studies of the thyroid follicle, Matrigel™ may offer some advantage over culture on plastic because there would be no distortion of the cells to adapt to the two-dimensional surface. For studies of thyroid cell function, it is difficult to assay multiple samples (e.g. dose–response curves) using Matrigel™, always assuming that cells thus cultured are indeed functional. For cell-signalling studies, Matrigel™ introduces another complexity because its chemical composition is ill-defined and the same problems arise as with the

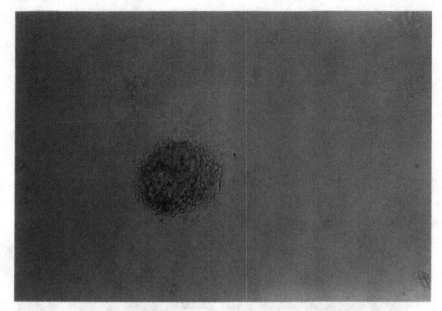

Fig. 7.5. A single human thyroid follicle, 6 h after plating, showing attachment to the plastic culture dish (×200 magnification).

use of serum. Finally, a further form of culture, involving cells plated onto permeable membranes offers the means to determine the polarity of secretion and receptor localisation. The thyroid cell is an excellent model for such studies because it is polarised (Chambard, Mauchamp & Chabaud, 1987). Although this model is difficult to set up, the polarity of secretion of thyroidal proteins and thyroid hormones can be determined using this technique as can the localisation of the iodide transporter and receptors for ligands.

Culture morphology

Within 6 h of plating, isolated thyroid follicles become adherent to the culture dish and start to spread over the surface (Fig. 7.5). Figure 7.6 shows photomicrographs of cells at (A) 1 and B,C,D) 5 days after isolation when they have been cultured with increasing concentrations of TSH. After 1 day the cells are attached to the dish and have spread out. After 5 days in the absence of TSH (Fig. 7.6 B), cells are confluent and flattened. TSH treatment causes the cells to retract their margins (Fig. 7.6 C). Although follicles can be seen in TSH-treated cultures (Fig. 7.6 D) under the light microscope,

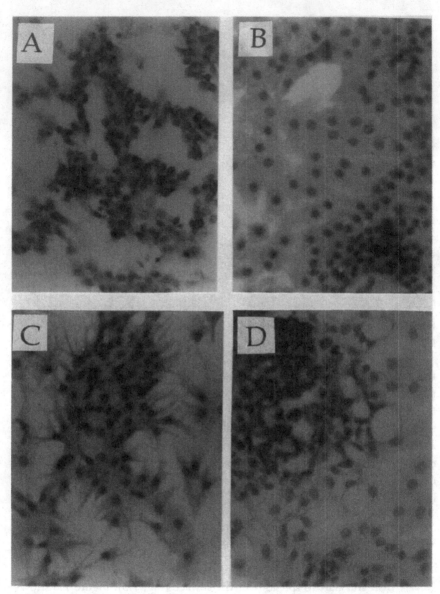

Fig. 7.6. Human thyroid follicular cells in monolayer. A, 1 day cultures; B, 5 day cultures not treated with TSH; C, as B but in the presence of 100 μU/ml TSH; D, as B but in the presence of 1 mU/ml TSH (×250 magnification).

the incidence of these is variable and not correlated with the rate of iodide metabolism. After 10 days of culture, the cell layer will be confluent, and cell growth will have decreased. Follicles will have formed and endogenous stores of thyroid hormones depleted. Human thyroid cells are more reluctant to form obvious follicles in culture than are sheep thyroid cells, where follicle formation is directly related to increasing TSH concentration. If follicles do form with the human cells they can be shown by immunostaining to contain thyroglobulin. Thyroid cell cultures obtained from multinodular goitres yield the most readily identifiable follicles, although it is unclear why this should be so. Follicles are three-dimensional, and with a phase contrast microscope it is possible to focus on the bottom and the top of the follicle.

Culture maintenance for studies of growth and function

Growth studies

Serum can be removed from the medium immediately after the cells have attached (1 day), although it is more usual not to change the medium until 3–5 days after plating. At this point the cells will be dividing but will not have reached confluence, and growth studies can be initiated. For such studies, cells should be washed at least twice in HBSS to remove serum and floating blood cells. Medium is then replaced with the agonist to be tested. Growth can be assayed by measurement of incorporation of [³H]-thymidine into DNA, although DNA measurements or cell counting unequivocally prove changes in cell number. The latter can be achieved with a Coulter counter or, more simply with a haemocytometer. Other more sensitive assays for measuring thyroid cell growth are available in the literature, and the method of choice will depend on the available instrumentation.

Functional studies

For assays of thyroid function, cells should be used 10–14 days after isolation. As a marker of thyroid cell function, uptake of radioactive iodide is uncomplicated. ^{131}I or ^{125}I can be used but the former isotope has a short half-life and a higher energy than ^{125}I. 0.1 μCi of ^{125}I is typically detected at 40–50% efficiency in gamma counters and yields approximately 10^5 cpm. This is easily and accurately assayed by the counter and cells on 2 cm² culture wells can take up between 10 and 80% of iodide given. The isotope should be added to the medium for periods between 1 and 16 h. The medium is then removed by suction into 5% sodium hypochlorite solution and the cell layer washed once with 1 ml ice-cold HBSS. It is important that the washing of the cell layer is performed rapidly because trapped iodide will rapidly re-equilibrate with fresh medium. Adding fresh medium containing the isotope rather than adding ^{125}I to the cell-conditioned medium generally results in a reduction in the amount of isotope taken up. Cells should then be removed from the culture plate for assay of their radioactivity by lysing them in 0.1M NaOH or detergents. Triton X-100 or other non-ionic detergents will lyse the plasma membrane but not the nuclei. Nuclei, which do not contain radioactivity, can be isolated in this way. Sodium dodecyl sulphate (SDS) (0.1%) will lyse the cells completely and can be used if an analysis of the structure of total cell proteins is required. The amount of iodide incorporated into proteins can be determined by precipitating the proteins in the cell layer with 10% trichloracetic acid. After incubation for 20 min in the cold the proteins should be pelleted by low speed centrifugation and counted in the gamma counter after removal of the supernatant. *Note*: Acid should be added to the cells in a fume chamber as acidification of solutions containing free radioactive ^{125}I produces volatile iodine.

If the ability of the cells to trap, but not organify iodide is to be examined, ^{99m}TcO$_4$ can be used. This large anion is trapped in a similar way to iodide, but is not organified. ^{125}I and ^{99m}TcO$_4$ can be readily distinguished by most gamma counters and use of both isotopes will give information on the activity of the iodide transporter as well as the ability of the cells to organify iodide, obviating the need to use peroxidase inhibitors. If ^{99m}TcO$_4$ is not available, cells can be incubated with methylmercaptoimidazole (10^{-4} M) to block thyroperoxidase. ^{125}I taken up will then be only attributable to trapped iodide. Variability in iodide uptake between preparations is probably an inherent feature of the surgical samples because the response of abattoir material is more predictable.

Other measures of thyroid cell function include assays of thyroid hormone release. For these studies the culture medium has to contain iodide. The optimal [NaI] for *de novo* synthesis is 100 nM, and higher concentrations are inhibitory to T3 and T4 synthesis. As for iodide uptake, considerable variations should be expected between successive cell preparations. Cells may conveniently be exposed to iodide for 16 h (e.g. overnight) and free T3 and free T4 determined in the medium by radioimmunoassay. Analysis of thyroid hormones in thyroglobulin in the cell layer can be performed after pronase digestion (10 mg/ml) in 0.1% SDS. The effect of SDS on the free T3 and T4 assays should be determined. These assays can be performed by clinical laboratories although this is an expensive route. An in-house assay can be set

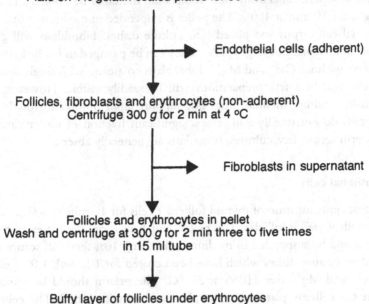

Trim tissue of fat and connective tissue
Chop using opposing scalpel blades to 2 mm³
Digest overnight at 16 °C in 0.2% collagenase until homogeneous
Filter through 250 µm filter with 50 - 100 ml HBSS
Centrifuge 300 g for 10 min at 4 °C
Plate on 1% gelatin-coated plates for 30 - 60 min

Endothelial cells (adherent)

Follicles, fibroblasts and erythrocytes (non-adherent)
Centrifuge 300 g for 2 min at 4 °C

Fibroblasts in supernatant

Follicles and erythrocytes in pellet
Wash and centrifuge at 300 g for 2 min three to five times
in 15 ml tube

Buffy layer of follicles under erythrocytes
Plate at 1 x 10⁴ - 5 x 10⁴ cells/cm²

Fig. 7.7. Summary of the isolation of the component cell types from human thyroid tissue.

up (Becks, Eggo & Burrow, 1987) or commercially available kits can be used (Eggo *et al.*, 1996). Human thyroid cells produce a disproportionate amount of T3, which may be due to high levels of deiodinase in the tissue.

Synthesis of thyroid-specific proteins or their mRNAs, e.g. thyroglobulin, thyroperoxidase, iodide transporter or TSH receptor can also be used as markers of thyroid function, using appropriate antisera or cDNAs. Assay of ^{125}I uptake (Eggo *et al.*, 1986) or cyclic AMP accumulation (Fransden & Krishna, 1976) in response to TSH are further quantitative measures of TSH receptor activity.

Isolation of other thyroid cell types

In addition to follicular cells, fibroblasts and vascular endothelial cells can be isolated from thyroid tissue digests, as summarised in Fig. 7.7.

Fibroblasts

The low speed centrifugation used to isolate the thyroid follicles (300 g for 2 min) does not effectively pellet the fibroblasts, which remain in suspension. The supernatants from these washes can therefore be pooled and centrifuged for 10 min at 400 g. The pellet is suspended in medium containing 10% fetal calf serum and plated into culture dishes. Fibroblasts will grow rapidly in this medium (Fig. 7.8 A). They can be passaged in 0.025% trypsin in HBSS without Ca^{2+} and Mg^{2+}. Fibroblasts co-isolate with single follicular cells, and in a few preparations will be readily visible. However, the serum-free culture medium does not support rapid fibroblast growth and such cells do not usually comprise a significant fraction of the culture. In long-term serum-free cultures, fibroblasts are generally absent.

Endothelial cells

The first centrifugation of thyroid follicular cells for 10 min at 300 g, yields single cells as well as follicles. This pellet, which also contains the endothelial cells, should be suspended in medium containing 10% fetal calf serum and plated on culture dishes which have been coated for 1 h with 1.0% gelatin in Ca^{2+} and Mg^{2+}-free HBSS at 37 °C. The gelatin should be removed before the cells are plated, and the plate washed with HBSS. The cells are incubated on the plates until the thyroid follicles start to attach. This can vary between 30 min and 1 h. At this point, the suspension of follicles is removed and the culture plate washed at least four times with HBSS. The thyroid follicles are centrifuged for 2 min at 300 g and washed at least four more times before plating as described above. The thyroid endothelial cells may then be cultured as described by Thornton, Mueller & Levine. (1983) (Fig. 7.8B). Further purification, to remove thyroid follicular cells and fibroblasts, will be required when the cells have grown up, and this may be achieved using fluorescent activated cell sorting (FACS) with Factor VIII antibody.

C-cells

The C-cells are a minor component of the normal thyroid gland, and it is unlikely that a sufficient quantity can be isolated by fluorescence-activated cell sorting (FACS) using antibody to calcitonin. Medullary carcinomas of the thyroid are predominantly derived from C-cells and can be used as a source of this cell type. Although immunostaining for calcitonin in thyroid

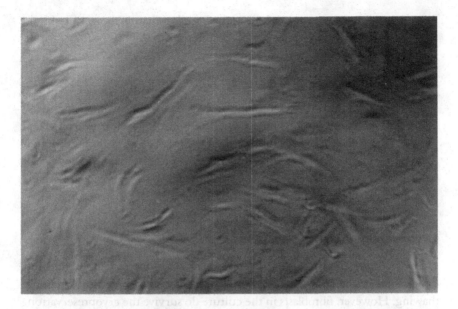

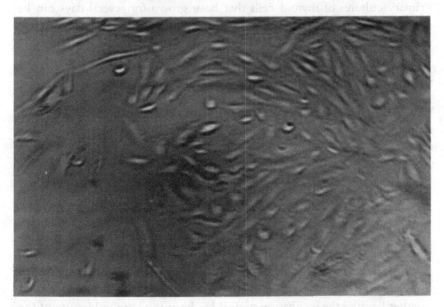

Fig. 7.8. Non-follicular cell types isolated from human thyroid tissue by centrifugation of supernatant after removal of thyroid follicles. top panel: fibroblasts, lower panel: endothelial cells (magnification ×100).

follicular cell preparations rarely reveals C-cell contamination, this should not be taken as definitive proof of their absence.

Longer-term culture and passage

Long-term cultures of thyroid follicular cells can be maintained for many passages, and these continue to grow in response to serum or epidermal growth factor (EGF) and respond to TSH (Errick *et al.*, 1986). Although long-term cultures do not respond to TSH as a mitogen, they do show TSH regulation of thyroglobulin synthesis. They do not, however, trap and organify iodide.

Cryopreservation

Using the standard procedures of freezing in 10% dimethyl sulphoxide and 10% serum in medium, freshly isolated thyroid follicles are not viable on thawing. However, fibroblasts in the culture do survive the cryopreservation. Primary cultures of thyroid cells that have grown for several days can be cryopreserved in liquid nitrogen, but the cells do not retain the ability to trap and organify iodide.

Troubleshooting

The isolation and culture of thyroid follicular cells is less problematical than those from many other endocrine tissues. However, occasional problems may arise, as detailed below.

1 *No viable cells obtained.* One possible reason for this is that the patient may have been treated with ablative doses of radioactive iodine prior to surgery. The preoperative treatment of the patient should therefore be checked prior to each cell isolation. Unless good liaison is established with the surgical staff, it is possible that tissue may have been frozen, or immersed in formalin prior to receipt in the laboratory. Poor viability of the cell preparation may also reflect damage during the collagenase digestion, either because the incubation period has been too long or because of the use of an unsuitable batch of the enzyme. The latter cause may be remedied by the prior testing of collagenase preparations from a number of manufacturers. Finally, cell viability may be compromised if one of the solutions to which the tissue or cells have been exposed is not isotonic.

2 *Cells do not attach.* This may be caused by a lack of attachment factors

within the culture medium. It may be possible to improve the rate and extent of attachment of newly seeded cells by increasing the serum content of the culture medium (e.g. to 5%). Occasional batches of culture plates may give less than optimal cell attachment. Check that they are, in fact, treated for tissue culture, and if the problem remains, try a different manufacturer's product.

3 *Contamination.* Thyroid tissue obtained from the operating theatre is rarely contaminated, but if this is suspected to be the case, the tissue may be sterilised by immersing briefly in 50–70% ethanol prior to processing. Contamination of cultures subsequent to plating out may reflect poor aseptic technique on the part of the operator, or contamination within the laboratory. Antibiotic supplements to the medium may be used to prevent or minimise this, although such use should not be seen as a substitute for good aseptic technique.

4 *No functional cells.* Isolated batches of collagenase may contain contaminants which, while not actually killing the cells, may impair function (e.g. through damaging cell adhesion molecules or TSH receptors). Optimal batches of collagenase should therefore be identified by batch testing, as described above. Some batches of serum can also impair thyroid cell function. If this is suspected, the cultures should be washed in HBSS, and fresh medium added. Loss of TSH responsiveness may indicate damage to the TSH receptor population during the cell isolation process; this may be checked by determining the response to cAMP analogues or adenylate cyclase stimulators such as forskolin, which raise cAMP independently of TSH receptor occupancy. However, it should not be assumed that a poor response to TSH invariably reflects cell damage; some batches of TSH have an inherently low bioactivity and if this is suspected, a fresh preparation should be tested. Finally, one of the major drawbacks associated with the use of diseased human thyroid tissue is that of possible unknown damage to the cells as a consequence of the disease process itself. This may, once again, give rise to a loss of thyroid cell function *in vitro*.

References

Al-alawi, N., Rose, D.W., Buckmaster, C., Ahn, N., Rapp, U., Meinkoth, J. & Feramisco, J. (1995). TSH-induced mitogenesis is *ras*-dependent but appears to bypass the *raf*-dependent cytoplasmic kinase cascade. *Mol. Cell. Biol.*, **15**, 1162–8.

Ambesi-Impiombato, F.S., Parks, L.A.M. & Coon, H.G. (1980). Culture of hormone-dependent functional epithelial cells from rat thyroids. *Proc. Natl Acad. Sci., USA*, **77**, 3455–60.

Becks, G.P., Eggo, M.C. & Burrow, G.N. (1987). Regulation of differentiated thyroid function by iodide: preferential inhibitory effect of excess iodide on thyroid hormone secretion in sheep thyroid cell cultures. *Endocrinology*, **120**, 2560–75.

Brown, R.S. (1995). Editorial: immunoglobulins affecting thyroid growth: a continuing controversy. *J. Clin. Endocrinol. Metab.*, **80**, 1506–8.

Chambard, M., Mauchamp, J. & Chabaud, O. (1987). Synthesis and apical and basolateral secretion of thyroglobulin by thyroid cell monolayers on permeable substrate: modulation by thyrotropin. *J. Cell Physiol.*, **133**, 37–45.

Dai, G., Levy, O. & Carrasco, M. (1996). Cloning and characterisation of the thyroidal iodide transporter. *Nature*, **379**, 458–60.

Davies, D.T. (1993). Assessment of rodent thyroid endocrinology. Advantages and pitfalls. *Comp. Haematol. Int.*, **3**, 142–52.

Divi, R.L. & Doerge, D.R. (1994). Mechanism-based inactivation of lactoperoxidase and thyroperoxidase by resorcinol derivatives. *Biochemistry*, **33**, 9668–74.

Drucker, D., Eggo, M.C., Salit, I.E. & Burrow, G.N. (1984). Ethionamide-induced goitrous hypothyroidism. *Ann. Int. Med.*, **100**, 837–9.

Eggo, M.C. (1993). Protein kinase C in the thyroid. *J. Endocrinol.*, **138**, 1–5.

Eggo, M.C., Bachrach, L.K., Fayet, G., Errick, J.E., Kudlow, J., Cohen, M.F. & Burrow, G.N (1984). The role of growth factors and serum on DNA synthesis and differentiation in thyroid cells in culture. *Mol. Cell. Endocrinol.*, **38**, 141–50.

Eggo, M.C., Bachrach, L.K., Mak, W.W. & Burrow, G.N. (1986). Disparate uptake of $^{99M}TcO_4$ and ^{125}I in thyroid cells in culture. *Horm. Metab. Res.* **18**, 167–76.

Eggo, M.C., King, W., Black, E.G.& Sheppard, M.C. (1996). Functional human and thyroid cells and their insulin-like growth factor binding proteins. *J. Clin. Endocrinol. Metab.*, **81**, 3056–62.

Errick, J.E., Ing, K.W.A., Eggo, M.C. & Burrow, G.N. (1986). Growth and differentiation in cultured human thyroid cells: effects of epidermal growth factor and thyrotropin. *In Vitro Cell. Develop. Biol.*, **22**, 28–36.

Fransden, E.K. & Krishna, G. (1976). A simple ultrasensitive method for the assay of cyclic AMP and cyclic GMP in tissues. *Life Sci*, **18**, 529–41.

Frohlich, E., Wahl, R. & Reutter, K. (1995). Basal lamina formation by porcine thyroid cells grown in collagen and laminin-deficient medium. *Histochem.J.*, **27**, 602–8.

Greenspan, F.S. & Strewler, G.J. (1997). *Basic and Clinical Endocrinology*, 5th edn. Appleton & Lange.

Gupta A., Eggo, M.C., Uetrecht, J.P., Cribb, A.E., Daneman, D., Rieder, M.J., Shaw, N.H., Cannon, M. & Speilberg, S.P. (1992). Drug-induced hypothyroidism. *Clin. Pharmacol. Therap.*, **51**, 56–67.

Hill, R.N., Erdreich, L.S., Paynter, O.E., Roberts, P.A., Rosenthal, S.L. & Wilkenson, C.F. (1989). Thyroid follicular cell carcinogenesis. *Fund Appl. Toxicol.*, **12**, 629–77.

Maenhaut, C., Lefort, A., Libert, F., Parmentier, M., Raspe, E., Roger, P., Corvilain, B., Laurent, E., Reuse, S., Mockel, J., Lamy, F., Van Sande, J. & Dumont, J.E.

(1990). Function, proliferation and differentiation of the dog and human thyrocyte. (1989) *Horm. Metab. Res.*, **23**, 51–61.

Nagataki S (1975). Effect of excess quantities of iodide. In *Handbook of Physiology, vol. III, Thyroid*, ed. R.O. Greep & E.B. Astwood, pp. 329–44 American Physiological Society.

Reid, J.D., Chang-Hyun, C. & Oldroyd, N.O. (1987). Calcium oxalate crystals in the thyroid. *Am. J. Clin. Pathol.*, **87**, 443–54.

Roger, P., Taton, M., Van Sande, J. & Dumont, J.E. (1989). Mitogenic effects of thyroptropin and cAMP in differentiated, normal human thyroid cells. *J. Clin. Endocrinol. Metab.*, **66**, 1158–65.

Studer, H. & Derwahl, M. (1995). Mechanisms of non neoplastic endocrine neoplasia. *Endocr. Rev.*, **16**, 411–26.

Thomas, G.A., Neonakis, E., Davies, H.G., Wheeler, W.H. & Williams, E.D. (1994). Synthesis and storage in rat thyroid C-cells. *J. Histochem. Cytochem.*, **42**, 1055–60.

Thornton, S.C., Mueller, S.N. & Levine, E.M. (1983). Human endothelial cells: use of heparin in cloning and long-term serial cultivation. *Science*, **222**, 623–5.

Wheeler, M.H. (1994). The solitary nodule. In *Diseases of the Thyroid*, ed. W.H. Wheeler & J.H. Lazarus, pp. 231–44. London, UK: Chapman and Hall.

Yamada, T., Kajihara,K., Takemura, Y. & Onaya, T. (1975). Antithyroid compounds. In *Handbook of Physiology, vol III, Thyroid*, ed. R.O. Greep & E.B. Astwood, pp. 345–58. American Physiological Society.

8

Hypothalamic cells

Hilary E. Murray, Duncan McKenzie and Glenda E. Gillies

Introduction

The hypothalamus, which is located on the ventral surface of the dien-cephalon, is a complex brain region involved in many functions and, in many instances, acts as the final common pathway for the integration of signalling between higher brain centres and the periphery (Fig. 8.1). Thus its roles include regulation of pituitary hormone secretion (and hence growth, development, metabolism, reproduction, the response to stress, salt and water balance), modulation of autonomic nervous function (via brainstem connections) and influences on behaviour, memory and learning (via connections to regions such as the limbic system). In accord with this functional complexity, the cytoarchitecture of the hypothalamus is also complex, consisting of heterogeneous neuronal populations synthesising an extensive range of neuropeptides and/or neurotransmitters. There are, however, a number of anatomically discernible cell groups or nuclei (Fig. 8.1), many of which have defined functions. This has been demonstrated particularly for the 'neuroendocrine hypothalamus', a title relating to its role in endocrine homeostasis, and some of this information is summarised in Table 8.1.

Research applications

The hypothalamus is relatively inaccessible to *in vivo* investigations, the stress of which also activates many homeostatic mechanisms converging on the hypothalamus. Thus *in vitro* methods for studying hypothalamic function are important alternatives for investigating the endocrine hypothalamus (Buckingham & Gillies, 1992). Cell cultures, in particular, offer an opportunity to study the function and development of hypothalamic cells (both neurones and glia) directly in a controlled, chemically defined medium which is free of serum and thus eliminates the interference of serum com-

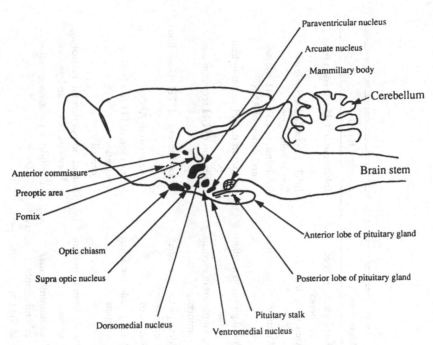

Fig. 8.1. Sagittal section of the brain highlighting some major hypothalamic nuclei and the anatomical relationship between the hypothalamus and the pituitary gland.

ponents which are essentially non-physiological (Clarke & Gillies, 1988). As the cultures consist of many cell types, a number of morphological, functional, and pharmacological measures can be used to focus on a particular cell group. Thus, for example, peptidergic neuronal populations can be studied using immunoassays to measure peptide release (basal and stimulated) and content (Clarke, Lowry & Gillies, 1987; Davidson & Gillies, 1993). This approach has been used to investigate the ontogeny of neurotransmitter control of neuropeptide secretion (Clarke & Gillies, 1988). For example, acetylcholine and 5-hydroxytryptamine (5-HT) have well-established abilities to activate the hypothalamo-pituitary-adrenal axis in the mature animal. In hypothalamic cell cultures these neurotransmitters are initially relatively inactive at time points corresponding to the first week after birth, but they gain dramatic increases in potency as the cultures mature. The period of relative insensitivity *in vitro* corresponds to the physiologically significant period of hyporesponsiveness of the stress axis in the new-born animal and the culture studies thus reveal a potential mechanism involving the central hypothalamic control of this axis in neuroendocrine development. Similarly, developmental changes in the pharmacology of gamma-amino butyric acid

Table 8.1. *Major cell groups of the neuroendocrine hypothalamus*

Neurochemical identity	Principal location	Projections	Ascribed function
Somatostatin (SRIH)	Periventricular nucleus (PeVN)	Portal plexus in median eminence	Inhibition of GH secretion from a.p.
	Periventricular nucleus (PeVN) Ventromedial nucleus (VMN)	Arcuate nucleus	Inhibition of GHRH release Sexual behaviour
	Paraventricular nucleus (PVN)	Within PVN/brainstem	Inhibition of CRH-41 release/ modulation of ANS
Growth hormone releasing hormone (GHRH)	Arcuate nucleus (arc)	Portal plexus in median eminence	Inhibition of GH secretion from a.p.
	Ventromedial nucleus (VMN)	Portal plexus in median eminence	Inhibition of GH secretion from a.p.
Corticotrophin releasing hormone (CRH-41)	Paraventricular nucleus (PVN)	Portal plexus in median eminence	Stimulation of ACTH release from a.p.
	Paraventricular nucleus (PVN)	Brainstem	Modulation of ANS
Gonadotrophin releasing hormone (GnRH)	Medial preoptic area (mPOA)	Portal plexus in median eminence	Regulation of LH and FSH secretion
	Scattered	? – central	? – central
Thyrotrophin releasing hormone (TRH)		Portal plexus in median eminence	Stimulation of thyroid stimulating hormone (TSH) and prolactin
Dopamine	Arcuate and periarcuate nucleus	Tuberoinfundibular/ tuberohypophyseal tracts to the portal plexus	Inhibition of prolactin secretion

Medial zona incerta	Anterior hypothalamus dorsal medial nucleus (DMN)	Modulation of hypothalamus-pituitary-gonadal axis
Periventricular nucleus (PeVN)		Influence on GH secretion (+/- depending on site of action)
GABA/Glutamate		
Diverse	Diverse (virtually all hypothalamic cells are responsive to GABA)	Complex modulation of neuroendocrine function
Vasopressin		
PVN and supraoptic nucleus (SON) (magnocellular division)	Posterior lobe of pituitary gland	Regulation of water balance and extracellular fluid volume (diuresis and vasoconstriction)
PVN (parvocellular division)	Portal plexus in median eminence	Regulation of ACTH release from a.p.
Oxytocin		
PVN, SON (magnocellular division)	Posterior lobe of pituitary gland	Parturition; milk ejection reflex

Note:
This list is not exhaustive and is continually growing. Other important neuropeptides include the opioid peptides, neurotensin (NT), cholycystokinin (CCK), vasoactive intestinal polypeptide (VIP). Many cell types have intra- and extra- hypothalamic projections with unknown functions. Abbreviations not defined in table: ANS, autonomic nervous system; a.p, anterior pituitary gland; ACTH, adrenocorticotrophic hormone; LH, luteinising hormone; FSH, follicle-stimulating hormone.

(GABA)-regulated somatostatin (SRIH) secretion support a role for GABAergic systems in regulating neuroendocrine development (Murray & Gillies, 1997). Such hypotheses can be investigated further by monitoring receptor or peptide expression as the cultures develop, using well-established ligand binding methods as well as molecular biological methods, including *in situ* hybridisation and reverse-transcription polymerase chain reaction, to monitor mRNA expression.

Factors which influence the development and function of hypothalamic neurotransmitter systems can also be followed by monitoring release and content of the endogenously synthesised transmitter. For example, high performance liquid chromatography has been used for the detection of dopamine (Murray & Gillies, 1993) and GABA (Murray & Gillies, 1997). In addition, the uptake kinetics and release of radioactively labelled neurotransmitters, e.g. tritiated GABA or dopamine, can be followed as the neurones develop (Murray & Gillies, 1993).

The morphological development of hypothalamic neuronal populations may also be studied after immunocytochemical identification of specific neuropeptides, neurotransmitters or synthesising enzymes, e.g. tyrosine hydroxylase for dopamine (Murray & Gillies, 1993) and glutamic acid decarboxylase for GABA. Figure 8.2, for example, illustrates the immunocytochemical localisation of tyrosine-hydroxylase-positive neurones in primary cultures of fetal rat hypothalamic cells. Morphometric analysis of cell numbers, neurite branching and length provides further power as a method of analysis.

Genetic vs. epigenetic influences on sexual dimorphism in the neuroendocrine system

A particular feature of the endocrine system is its sexual dimorphism which arises, in part, from the actions of gonadal steroids during a critical perinatal period to programme, irreversibly, the central circuitry governing neuroendocrine function in the mature animal. The cellular and molecular mechanisms underlying these processes remain poorly defined and recent studies, employing primary hypothalamic cell cultures from male and female animals (mice and rats) are beginning to provide insight into the genetic and epigenetic influences which predict the 'maleness' and 'femaleness' of the brain and the neuroendocrine system in particular (Reisert & Pilgrim, 1991). Our own studies reveal a particular sensitivity of certain hypothalamic populations to prevailing levels of estrogen and testosterone, which depends on the male/female genotype (Christian *et al.*, 1997). Certain chemicals from

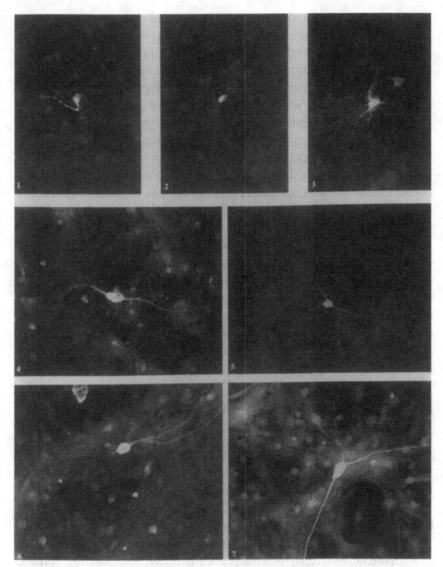

Fig. 8.2. Immunocytochemical localisation of tyrosine hydroxylase (TH) immunopositive neurones in primary cultures of foetal rat hypothalamic cells. TH is the rate limiting enzyme in the synthesis of dopamine, noradrenaline and adrenaline. As the cell bodies of adrenergic and nor-adrenergic neurones are not found in the hypothalamus, these cells represent dopaminergic neurones. Three distinct subtypes are seen on day 3 *in vitro* (1–3) as determined by their uni-, bi- and multipolar appearance. By days 7 (4, 5) and 14 (6, 7) only the bi- and multipolar cells are seen. (Taken, by permission from Murray & Gillies, 1993).

both industrial and natural sources, which are found in our environment and act as agonists or antagonists at the gonadal steroid receptors, produce equally dramatic effects on developing hypothalamic neurones (Christian *et al.*, 1997). The use of cultured cells to screen pollutants, which may disrupt developing central neurones, might thus prove effective as an indication of xenobiotic compounds which merit further toxicological assessment.

The following protocols will describe first, the preparation of primary cultures from isolated rat hypothalamic cell suspensions and second, the preparation of slice cultures. The latter method is a relatively newer technique which has the advantage that a three-dimensional cytoarchitecture is retained to a certain extent, but with the former method the cells re-establish themselves *in vitro* more quickly in a serum-free environment and some degree of histotypic reorganisation *in vitro* has been reported using this method (Hemmendinger *et al.*, 1981). The slice culture method is included here as it lends itself particularly well to studies of cell interactions in co-culture. For example, slices taken from two brain areas (e.g. the hypothalamus and the hippocampus) which are normally connected in the intact animal, may be cultured together on the same membrane patch. As the co-culture develops, connections may be established between the two parts, and the factors influencing development and regeneration of tissue structure may thus be investigated.

Preparation of dissociated hypothalamic cell cultures

Materials and media
Sterile petri dishes (150 mm diameter; Life Technologies)
Sterile culture well plates (35 mm and 14 mm diameter; Life Technologies)
Sterile saline: 0.9% NaCl supplemented with penicillin (100 U/ml, ICN) and streptomycin (100 μg/ml, ICN).
Collection buffer: 0.1 % w/v bovine serum albumin (BSA Fraction V, Sigma) 20 mM HEPES (ICN);100 U/ml penicillin (ICN); 100 μg/ml streptomycin (ICN); 100 ml Hank's Balanced Salt Solution (HBSS, Ca^{2+}/Mg^{2+} free; Life Technologies).
Enzyme solution: 0.4% w/v BSA; 100 U/ml penicillin; 100 μg/ml streptomycin; 20 mM HEPES (ICN); 0.2 mg/ml neutral protease (Dispase, Peninsula Laboratories); 0.5 mg/ml crude DNase I (Sigma); 40 ml HBSS (Ca $^{2+}/Mg^{2+}$ free).
DNase solution: 0.4% w/v BSA; 100 U/ml penicillin; 100 μg/ml streptomycin; 20 mM HEPES; 0.5 mg/ml DNase (Sigma); 20 ml (HBSS, with Ca^{2+}/Mg^{2+} ; Life Technologies).

Defined medium: 50 ml Dulbecco's Modified Eagle's Medium (Life Technologies); 50 ml Ham's F12 Nutrient Medium (Life Technologies); 100 U/ml penicillin; 100 μg/ml streptomycin; 0.5 μg/ml fungizone (ICN); 5 μg/ml insulin (Sigma); 100 μg/ml human transferrin (Sigma); 3×10^{-8}M selenium; 1×10^{-4}M putrescine; 2×10^{-8}M progesterone* (Sigma); 1×10^{-12}M 17β-estradiol* (Sigma); 1×10^{-9}M tri-iodithyronine* (Sigma).

Note: Fresh stock solutions of defined medium constituents are prepared monthly, aliquoted and stored at −20 °C until use.* Concentrated stocks of steroids are made up in ethanol; diluted stocks are stored in siliconised glass bottles (to prevent non-specific adsorption to the glass surface) at −20 °C.

Tissue dissection

Protocol

1 Female Wistar rats, time-mated to be 18 days pregnant on the day of experiment, are placed in a clean area for the removal of the fetuses. Depending on the purpose of the experiment, tissue could also be collected at earlier developmental stages as well as later, up to about 1 week postnatally in the rat, before survival is severely compromised. Place the mothers one at a time into an anaesthetising box. Following inhalation of diethyl ether or halothane vapours, dams rapidly become anaesthetised and aseptic removal of the fetuses may be achieved by Caesarean section under continual anaesthesia. (Following this procedure, the mothers should be killed according to approved local rules for animal handling.)

2 Place the uterus containing the fetuses into a sterile petri dish containing 5 ml of sterile saline supplemented with antibiotics.

3 Transfer the tissue to the tissue culture laboratory where the fetuses are excised, and decapitated.

If gender specific cultures are required the pups should be sexed prior to decapitation by determining the ano-genital distance (greater in the male than female) and/or by visualisation of the testicular artery under dark field microscopy (Kauffman, 1992).

4 Transfer each head to a sterile petri dish containing a saline-soaked filter paper. Curved forceps may be used to hold the head in position, while an incision is made from the base of the skull to the snout with iridectomy scissors to expose the dorsal surface of the brain. Use curved forceps to lift the brain at the point of the frontal lobes and invert it to expose the ventral surface.

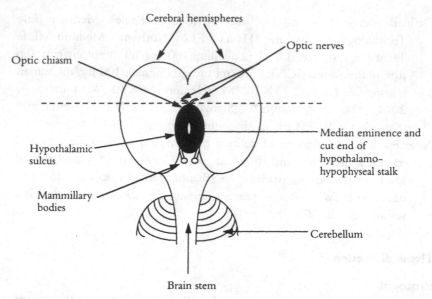

Fig. 8.3. Diagrammatic representation of the foetal hypothalamus and surrounding structures.

5 The hypothalamus of the fetal brain is a prominent structure, (2–3 mm across) and is removed using fine curved watchmakers' forceps. A 'pinch' of tissue that encloses the 'hypophysiotrophic area' bordered by the hypothalamic sulci laterally, the mammillary bodies caudally, and the optic chiasm rostrally, is dissected out (approximately 1 mg wet weight; see Fig. 8.3).

6 Place the dissected hypothalami in a sterile petri dish containing collection buffer. The tissue gathered from a number of pregnant rats is pooled for each set of investigations. Ideally two workers will be involved in this stage of the procedure and 100 hypothalami may be collected in 90 min (from approximately ten mothers).

Cell dispersal

Handling of cell suspensions from this stage onwards should be performed within a Class II Microbiological Safety Cabinet.

Protocol

1 Transfer the pooled hypothalami to a sterile 30 ml Universal Container and pellet them by centrifugation at 900 rpm for 3 min.

2 Resuspend the pellet in 2 ml enzyme solution pre-warmed to 37 °C and

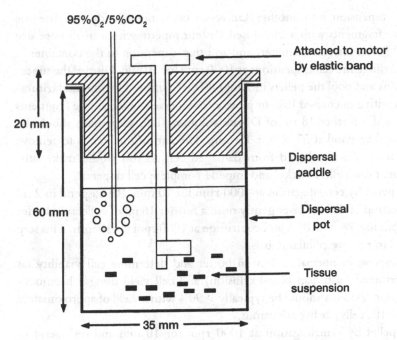

95%O_2/5%CO_2

Attached to motor
by elastic band

20 mm

Dispersal
paddle

60 mm

Dispersal
pot

Tissue
suspension

◄——— 35 mm ———►

Fig. 8.4. Dispersal apparatus for the isolation of hypothalamic cells from whole tissue.

gently triturate using a sterile, siliconised Pasteur pipette with an internal bore size of approximately 1.5 mm (prepared by careful flaming in a bunsen burner flame prior to sterilisation). Add a further 18 ml of enzyme solution and transfer the cell suspension to the dispersal apparatus (Fig. 8.4). This is then placed in a water bath (37 °C) containing distilled water and a mild disinfectant (Roccal). Gently gas the contents of the dispersal apparatus with 5% CO_2/95% O_2 and mechanically agitate using a teflon paddle, driven by an electric motor and set to rotate at approximately 200 rpm.

3 After 20 min, allow the larger tissue fragments to settle at the bottom of the dispersal pot and decant the supernatant containing the cell suspension into a sterile Universal container. This should then be placed in an incubator set at 37 °C. Resuspend the larger fragments in 2 ml enzyme solution and further triturate with a siliconised Pasteur pipette with a smaller bore size than that used initially (approximately 1.0 mm internal diameter).

4 Add a further 18 ml of enzyme solution to these fragments and return the dispersal apparatus to the water bath for a further 15 min under the same conditions as for the first incubation.

5 Once again, allow any remaining tissue fragments to settle and decant the

cell suspension into another Universal container. Gently triturate the tissue fragments with a siliconised Pasteur pipette with a small bore size (<0.5 mm internal diameter) and add this suspension to the container.

6 Centrifuge the cell suspensions at 1000 rpm for 10 min. Discard the supernatants and pool the pellets in 2 ml DNase solution. Use minimal trituration with a siliconised Pasteur pipette to disperse any remaining fragments and add a further 18 ml of DNase solution. This suspension should be allowed to stand at 37 °C for 15 min. The aim of this step is to remove the nucleic acid released from damaged cells as this would make subsequent cell pellets 'sticky' and impede complete cell dispersal.

7 Re-pellet by centrifugation at 1000 rpm for 10 min. Re-suspend in 2 ml collection buffer and layer gently onto a further 16 ml of collection buffer containing 4% w/v BSA and centrifuge at 1000 rpm for 10 min. This step helps to remove cellular debris.

8 Resuspend in normal collection buffer and determine cell viability (as determined by trypan blue exclusion) and cell yield using a haemocytometer. Viability should be typically >90% with a yield of approximately 0.8×10^6 cells per hypothalamus.

9 Re-pellet by centrifugation at 1000 rpm for 10 min and resuspend in defined culture medium to give a cell density of approximately 2.5×10^6 cells/ml.

Cell plating and culture maintenance

Protocol

1 While continually swirling the cell suspension, gently transfer the appropriate volume to the culture vessels. If the production of endogenously synthesised neuropeptide or neurotransmitter is to be followed, 1 ml of cell suspension is plated into 35 mm diameter plastic culture dishes. Cell attachment is improved by pretreating the dishes with filter-sterilised poly-L-lysine (10 μg/ml sterile water) for 30 min at room temperature. This is followed by rinsing of the dishes with sterile water to remove all traces of poly-L-lysine. Attachment and survival may also be improved by a brief incubation of the dishes with a serum-supplemented medium which is completely removed before the cells are added. Each dish is swirled gently to ensure even layering of the cells. If the uptake and release of radioactively labelled neurotransmitters are to be studied, 180 μl of cell suspension should be plated into 15 mm diameter plastic tissue culture wells, treated as above. For histological or morphological examination of the cultures, cells are plated onto sterile 13 mm diameter glass coverslips which have been coated previously with poly-L-lysine (100 μg/ml over-

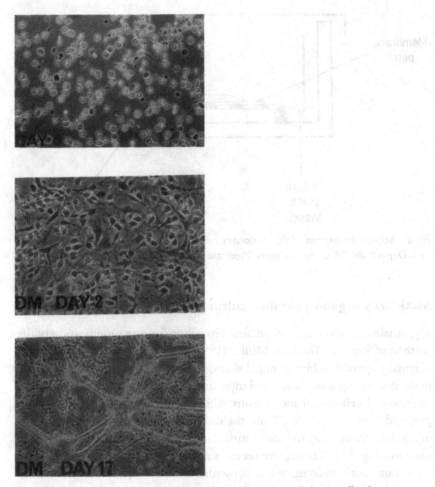

Fig. 8.5. Phase contrast photomicrograph of primary hypothalamic cell cultures maintained in defined medium for 0 (upper panel), 2 (centre panel) and 17 days (lower panel).

night, followed by extensive rinsing in sterile water and saline). These coverslips fit comfortably into the 15 mm diameter multiwell plates which makes manipulation of the cells easier.

2 Maintain the plated cells in an incubator at 37°C in 5% CO_2 / 95% air and 100% humidity.

3 Change the culture medium for the first time 4 days after plating and every third day from then onwards.

Figure 8.5 shows the morphological appearance of primary hypothalamic cultures immediately after plating, and following 2 and 17 days of culture.

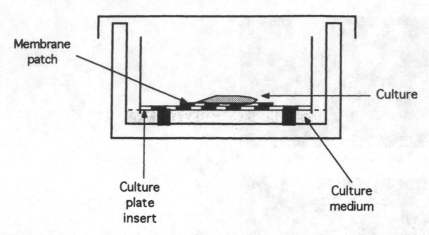

Fig. 8.6. Schematic diagram of the stationary culture apparatus (illustration provided by S. Duport and P. Correges, Centre Medicale Universitaire, Geneva, Switzerland).

Stationary organotypic slice cultures of rat hypothalamus

Hypothalamic slices can be maintained in culture using a modification of the method of Stoppini, Buchs & Muller (1991). Using this method, which was originally applied to hippocampal slices, the culture exhibits many features of the developing tissue *in vivo* and most importantly, maintains certain three-dimensional cell-to-cell interactions which could be disrupted in cultures prepared from isolated cells. Thus, the slice culture rests on the upper surface of a porous membrane, the lower surface of which is in contact with culture medium (Fig. 8.6). The culture receives nutrients and moisture through the membrane, but is in constant contact with a humidified, 5% carbon dioxide-enriched atmosphere.

Materials, media and equipment
Dissecting medium: 100 ml MEM (Gibco) supplemented with 2 mM L-glutamine (Gibco); 400 U/ ml penicillin – 400 μg/ml streptomycin (Gibco).

Culture medium: 30 ml dissecting medium plus 10 ml horse serum (Gibco).

Sterile six-well flat bottomed tissue culture plates (35 mm, Falcon).

Sterile culture plate inserts (Millipore).

Biopore™ membrane (Millipore) cut into patches of dimension ~8 mm and made aseptic by soaking overnight in ethanol and drying under ultra-violet irradiation in a laminar flow cabinet.

Autoclaved or aseptic dissection instruments and glassware.

One or two gassed (5% carbon dioxide) and humidified incubators.
Aseptic working area (e.g. laminar flow cabinet with ultraviolet lamp)
McIlwain tissue chopper (Mickle Laboratory Engineering Co.).

Protocol

1 Prepare a culture plate: put 1 ml of culture medium and a culture plate
   insert into each well; place two or three membrane patches onto each
   culture plate insert; the membrane patches and the membrane in the
   culture plate should be wetted by the culture medium, and become transparent. Place the culture plate in the incubator at 36 °C.
2 Decapitate a 5–7-day-old rat and remove the brain to a petri dish containing 3 ml of dissection medium. Transfer the petri dish containing the
   brain to the aseptic working area.
3 Slice the brain at the level of the optic chiasm (marked with a dotted line
   in Fig. 8.3); retain the portion containing the hypothalamus, keeping the
   tissue moist with dissecting medium.
4 Lay the block of tissue on the stage of the McIlwain chopper, aspirate any
   excess medium and cut slices 300 µm thick. Return the cut tissue to the
   dissecting medium, and select the slices containing the required nucleus
   under a dissecting microscope. Trim the selected slices and place them on
   the membrane patches in the culture plate, one culture to each patch of
   membrane.

The culture plate inserts are sterile when first used, but are routinely re-used
by soaking in 70% ethanol and then treating as the membrane patches to
achieve asepsis. They should be discarded if they show signs of deterioration,
or if they have supported any fungal infection.

Maintenance of the cultures

Cultures are maintained in a humidified, carbon dioxide (5%)–enriched
atmosphere at 36 °C, and the culture medium changed after the first 2 days,
then subsequently every 3–4 days. The cultures flatten and clear (i.e. become
transparent) after 7–10 days, and adhere to the membrane patch. Cultures
can be maintained for at least 4 weeks, although they are normally used
before this time. If two incubators are available one can be set at 37 °C and
the second at 34 °C. The cultures should be maintained for 2–4 days in the
warmer incubator, and then transferred to the cooler incubator for the
remainder of the incubation. Using two incubators in this way gives cultures
which clear more quickly, and are more likely to clear completely. The

culture medium and dissecting medium are stored at 4 °C, and used for no more than 7 days (culture medium) or 14 days (dissecting medium).

Note: Hypothalamic slice cultures are more difficult to maintain using this method than hippocampal slices. Hippocampal cultures are typically prepared using 400 μm thick slices. Hypothalamic cultures prepared from 400 μm slices tend to develop brown patches of necrotic tissue, and do not flatten and clear. This problem can be overcome by using thinner slices. Other brain areas may present similar difficulty, and slice cultures of spinal cord are therefore also prepared from 300 μm slices.

The viability of slice cultures has, to date, been verified by a variety of methods:

(1) *Electrophysiology*, e.g. single unit extracellular recordings have shown the presence of spontanaeously active neurones in the region of the hypothalamic paraventricular nucleus (PVN). Some of the neurones have distinctive firing patterns that are also seen in acute slice preparations.

(2) *Immunocytochemistry*, e.g. nuclei immunoreactive for CRH and oxytocin can clearly be seen in the region of the PVN.

(3) *Hormone release*, e.g. stimulation of slice cultures with 56 mM K^+ evokes release of the hormone somatostatin (SRIH).

(4) *Electron microscopy*, e.g. electron micrographs of hypothalamic slice cultures show the presence of neurones, glia and synapses.

Because the cultures adhere to a patch of porous membrane, they are easy to handle without risk of damage, which offers particular advantages. Thus the culture can be moved (e.g. to transfer it from a culture dish to an electrophysiological recording chamber) simply by picking up the membrane patch with a pair of forceps. This bypasses the step of dissolving the plasma clot or collagen matrix used in some slice culture methodology, or indeed of cutting the culture free from its cover slip, as is often necessary with organotypic cultures prepared by the roller tube method. It is also possible to carry out immunocytochemistry on cultures without removing them from the membrane patches, as the patch is transparent when wetted, although care is required when mounting the stained culture for long-term storage, as some mounting media can turn the membrane opaque. For experiments aimed at measuring the release of hormones or neurotransmitters, two to four cultures can be prepared in a single culture plate insert, then for each wash, the insert complete with its cultures can be picked up and moved to a well containing the next wash. Again, this technique reduces the risk of physical damage to the culture when changing solution. Despite the fact that the hormones or neurotransmitters released from the cultures have to diffuse through both the membrane patch and the membrane of the culture plate insert, satisfactory results have been obtained with this sort of experiment. To prepare the culture for electron microscopy, the culture can again remain attached to the membrane patch when it is fixed, embedded and sectioned.

Troubleshooting

As in all cell culture techniques, the most obvious cause for failure is bacterial and fungal infection. This can be readily avoided by strict adherence to aseptic procedure and good laboratory practice. All dissecting equipment should be heat sterilised or autoclaved and high quality sterile consumables (including pipette tips) must be used at all times. Bottles of medium should be stored at the recommended temperature and, once opened, should be handled with due care to prevent media contamination. As many aliquots of the additives for the defined medium have to be stored, it is wise to keep records of the date at which fresh aliquots are prepared and also of the batch or lot numbers of chemicals and serum. Thus, it can be checked whether a problem first appeared with the use of a new batch of a particular ingredient. As hypothalamic cells are very sensitive to fluctuations in pH, CO_2 and temperature the reliability of these readings should be checked regularly. If the action of estrogen on neurones in culture is being investigated, phenol red should be excluded from the medium as it is weakly estrogenic.

If the slice cultures are consistently unhealthy (e.g. brown patches of necrotic tissue are seen), reduce the number of cultures on each cell culture insert (two cultures per insert are typically used), and take more care to avoid damage to the tissue during dissection (this should improve considerably with practice).

References

Buckingham, J.C. & Gillies, G.E. (1992). In *Endocrine Toxicology*, vol. 1, ed. C.K. Atterwill & J.D. Flack, pp. 83–114. Cambridge, UK: Cambridge University Press.

Christian, M.C., Murray, H.E., Jones, D. & Gillies, G.E. (1997). Direct evidence for an influence of environmental oestrogens on developing rat hypothalamic dopaminergic neurones in culture. *J. Endocr.*, **152** (suppl) OC39.

Clarke, M.J.O. & Gillies, G.E. (1988). Comparison of peptide release from rat fetal hypothalamic neurones cultured in defined medium and serum-containing medium. *J. Endocr.*, **116**, 349–56.

Clarke, M.J.O., Lowry, P.J. & Gillies, G.E. (1987). Assessment of corticotrophin releasing factor, vassopressin and somatostatin secretion by fetal hypothalamic neurones in culture. *Neuroendocrinology,* **46**, 147–54.

Davidson, K. & Gillies, G.E. (1993). Neuronal vs glial somatostatin in the hypothalamus: a cell culture study of the ontogenesis of cellular location, content and release. *Brain Res.*, **624**, 75–84.

Gillies, G.E. (1997). Somatostatin: the neuroendocrine story. *TIPS,* **18**, 87–94.

Gillies, G.E., Davidson, K., Murray, H.E. & Christian, M.C. (1996). Gonadal steroids and hypothalamic neuronal development. *J. Physiol.,* **494P**, S10–S11.

Hemmendinger, L.M., Garber, B.B., Hoffman, P.C. & Heller, A. (1981). Target neuron-specific process formation by embyronic mesencephalic dopamine neurons *in vitro. Proc. Nat. Acad. Sci.,* USA, **78**, 1264–8.

Kauffman, M.H. (1992). *The Atlas of Mouse Development.* London: Academic Press.

Murray, H.E. & Gillies, G.E. (1993). Investigation of the ontogenetic patterns of hypothalamic dopaminergic neurone morphology and function *in vitro. J. Endocr.,* **139**, 403–14.

Murray, H.E. & Gillies, G.E. (1997). Differential effects of neuroactive steroids on somatostatin and dopamine secretion from primary hypothalamic cell cultures. *J. Neuroendocrinol.,* **9**, 287–95.

Reisert, I. & Pilgrim, C. (1991). Sexual differentiation of monoaminergic neurones-genetic or epigenetic? *TINS,* **14**, 468–73.

Stoppini, L, Buchs P.A. & Muller, D (1991). A simple method of organotypic cultures of nervous tissue. *J. Neurosci. Meth.,* **37**, 173–82.

Index

Page numbers in *italic* type refer to illustrations.

acetylcholine 135
acidophils 38
adrenal chromaffin cells 74
adrenal fasciculata–reticularis cells
  cortisol secretion 75
  isolation 77–9
  separation from granulosa cells 82
adrenal gland
  bovine
    dissection 77–8, *78*, 81–2
    location 77
    source 76
  cortex 74
    zona fasciculata 74
    zona glomerulosa 74
    zona reticularis 74
    human 88–91
      isolation and culture 89–90
      source 89
  medulla 74, 82–6
  rat
    dissection 87
    source 87–8
adrenal glomerulosa cells
  bovine, aldosterone secretion 82
  rat
    aldosterone secretion 88
    corticosterone secretion 88
    isolation 87–8
    primary culture 87–8;
      maintenance 88

adrenal reticularis–fasciculata cells 79,
      85
  purification on Percoll 79
adrenocortical cells
  bovine
    aldosterone secretion 75
    primary culture 75–81;
      maintenance 79–80
  human
    fetal 88
    primary culture 88–91
    source 89
  size-fractionation 82
adrenocorticotrophic hormone (ACTH)
      38, 74, 75
adrenomedullary cells
  catecholamine secretion 82–3, 86
  identification 83
  isolation 83–4
  primary culture 82–6
  purification 86
aldosterone, synthesis by adrenal
      glomerulosa cells 75, 81, 82
androgens, production by thecal cells
      24, 74
androstenedione 24, 25, 36
androsterone 25
angiotensin II 74, 75, 79
  effect on aldosterone secretion 75
anterior pituitary cells
  clusters 48–50

anterior pituitary cells (*cont.*)
   response to secretagogues 48
   ultrastructure 48
  culture morphology 49
  cytosolic Ca^{2+} measurement 53–7, *58*
  gradient centrifugation, discontinuous
    51–2
  hormone release 39–40, 45–50
  identification 53
  isolation 41–4
  lactotrophs
    perifusion 46–50, 53–6
    prolactin release from *58*, 59–60
    purification and enrichment 51–2
  somatotrophs 53
  thyrotrophs 53
anterior pituitary gland
  collection 41
  hormone secretion 38
antioxidants 80
ascorbic acid 59–60, 75, 80, 119–20
atresia 26, 29

basophils 38
Bio-gel™, cell perifusion 46–50
Biopore™ membrane 146
Bouin's solution 100

Ca^{2+}, cytosolic
  and catecholamine secretion 83
  and insulin release 63
  and parathyroid hormone secretion
    16, 17
  and prolactin secretion *58*
  in pancreatic ß cells *64*
  in parathyroid cells 9
  measurement in anterior pituitary cells
    53–7, *58*
calcitonin 128, 130
catecholamines, secretion from adrenal
    cells 82, 83, 86
chorionic gonadotrophin (CG) 94
chromaffin cells *see* adrenomedullary
    cells
collagen
  as coating for culture plates 122
  type I, rat tail 101
  type IV 122

collagenase 6, 7, 8–10, 13, 17–8, 43,
    68, 69, 72, 78, 84, 87, 90, 91,
    96–7, 100, 106–7, 116–17, 131
  crude 8, 13, 17
  purified 8–9, 13, 17, 18
  toxicity to cells 131
  type I 7, 8–9, 17, 18, 87, 117
  type II 78, 84, 90
  variation between batches 72, 78, 84,
    90, 106–7, 116–17, 130
corpus luteum 24
corticotrophin releasing hormone *136*
cortisol
  secretion from adrenal
        fasciculata–reticularis cells 75,
        82
  as supplement to culture medium 120
Coulter counter 79, 98, 125
cyclic AMP 51, 71, 75, 109, 113, 127,
    131
cyclic GMP 51
cytochrome P450
  aromatase 20
  C17–20 lyase 20
  cholesterol side-chain cleavage 20
cytosine arabinoside 86

DEAE Sephadex, substrate for cell
    attachment 48
dehydroepiandrostenedione 88
diacylglycerol 50
digestion cocktails
  optimisation of 16–17
  preparation 8–9
Dispase 117, 140
DNA
  effect on tissue dispersion 44, 91
  measurement 125
DNase 6, 10, 44, 78, 84, 87, 90, 91, 97,
    100, 140, 144
  type I 78, 84, 87, 90, 97, 100, 140
dopamine
  effect on prolactin release 59–60
  in neuroendocrine hypothalamus *136*

elastase 8, 13, 17, 18
endocrine pancreas, constutuent cell types
    62

endothelial cells   119, *127*, 128
entactin   122
epidermal growth factor (EGF)   75, 130
estradiol   20, 22
   production in ovarian follicle   20
estrogen, effects on hypothalamic cells
      138
estrogen priming   29, 60
   using diethylstilbestrol   29
ethylene-*bis*-dithiocarbamates   109

Factor VIII   112
   antibody to   128
fibroblast growth factor-2 (FGF-2)   75,
      80
fibroblasts   80, 83, 101, 113, 118, *127*,
      128, 130
   contamination by   101, 113
   identification   83
   overgrowth by   80
fibronectin   22, 36, 76, 80, 83, 85, 101
   coating for culture plates   36, 80, 101
   staining for fibroblasts   83
Fluo-3   53, 54
fluorescence-activated cell sorting (FACS)
      63, 128
follicle-stimulating hormone (FSH)   20,
      22, 38
forskolin   20, 22, 36, 131
Fura-2   53, *64*, 72

gamma-amino butyric acid   135, 138
   in neuroendocrine hypothalamus   *136*
gelatin   128
glibenclamide   67
glucocorticoids, synthesis by
      adrenocortical cells   74
glycyl-histidyl-lysyl acetate   120
goitrogens   109-10, 114, 122
   effect on thyroid tumourigenesis   110
   screening for   114
gonadotrophin releasing hormone   *136*
Graffian follicle   20
   morphology   *21*
growth hormone (GH)   38, 51
   secretion by somatotrophs   51
growth hormone releasing hormone
      136

haemolytic plaque assay   53, 67
heparan sulphate proteoglycan   122
hydroxyprogesterone, 20a   95
hydroxysteroid dehydrogenase, 3ß-   20,
      24, 99, 101, 103
5-hydroxytryptamine   135
hypoglycaemic sulphonylurea drugs   67
hypothalamic cells
   rat
      culture morphology   *139, 145*
      dopamine release   138
      glutamic acid decarboxylase release
         138
      isolation   141-4, *143*
      maintenance of cultures   144-5
      somatostatin release   138
      tyrosine hydroxylase expression
         138
hypothalamic paraventricular nucleus
      148
hypothalamic slice culture   140,
      146-9
   apparatus   *146*
   rat
      maintenance   147-9
      preparation   147
hypothalamus   *135, 142*
   foetal   *142*

Indo-1   53, 54, 56
insulin
   release, physiological stimulators of
      63
   as supplement to culture medium   11,
      35, 75, 80, 99, 100, 120
insulin-like growth factor binding
      proteins   113
insulin-like growth factor-I (IGF-I)   75,
      80, 113
iodide transporter   127
iodotyrosines   111
ionomycin   56-7
Islets of Langerhans   62
isobutylmethylxanthine   71

laminin   101, 122
   as coating for culture plates   101

Leydig cell cultures, contamination by
    fibroblasts 101
Leydig cells
  cryopreservation 106
  human 96–7, 100, 101–5
    culture morphology *102, 104*
    responsiveness *105*
    source 96
    testosterone secretion from *105,*
    106
  porcine 95, 97–100, 103, 105–6
    culture morphology *102*
    isolation and culture 97–100
    plating efficiency and
      characterisation 103, 105–6
  rat 95, 106

Lima bean trypsin inhibitor 41, 43, 44
low density lipoproteins, supplement to
    culture medium 80
luteinising hormone (LH) 20, 36, 38,
    94

Matrigel 122–3
McIlwain tissue chopper 147
metalloproteinases 116
mineralocorticoids 74

neurotransmitters 135, 148–9
non-insulin dependent diabetes mellitus
    63

oocyte
  morphology *21*
  retrieval 26
opioid peptides, secretion by
    adrenomedullary cells 82
orchidectomy 96
ovarian follicle 20, 26–30
ovarian granulosa cells
  characterisation 22
  cryopreservation 36
  culture morphology *23*
  estradiol production by *24*
  isolation 26–30
    by aspiration 28–9
    by follicle puncture 29–30

    by microdissection 27–8
  maintenance 34, 36
  response to FSH 36–7
  source 25–6
ovarian theca
  cells
    androgen production by *25*
    characterisation 22, 24–5
    cryopreservation 36
    culture morphology *23*
    isolation 22–6
    maintenance 34, 36
    microdissection 30–2
    Percoll gradient purification 32–4
      35, 37
  tissue
    externa 20, *21*, 31
    interna 20, *21*, 22, 31

oxytocin *137*

pancreas, rat *69*
pancreatic endocrine cells,
    cryopreservation for
    transplantation 71
pancreatic ß-cells
  heterogeneity 67
  morphology and characterisation
    65–7, *66*
  purification 63
  rat
    isolation 67–70
    microdissection 68–9
    response to glucose *64,* 67, 71
  sources 67–8
  transplantation 65
papain 8, 11
parathyroid cell culture
  monolayer 6, 7, 11, *12*
  organoid 7, 11–13, *14, 15, 16, 17*
  suspension 7
parathyroid hormone (PTH) 6
  secretion 6, 11, 13
parathyroid tissue
  bovine 6, *14*
    digestion 7–11
    heterogeneity 7

calf 9
digestion, effect on plasma membranes
    and receptors 8–9
human 6, 9
porcine 6
Percoll 28, 30, 31–4, *35*, 37, 52, 79, 84,
    88, 90, 91–2, 98–100
PhenoCen 9, 18
phenol red 59–60, 84, 149
estrogenic effect of 59–60, 149
pituitary gland
anatomy 38–9
anterior 39, *42*
    rat: collection and dissection 41,
        42; culture morphology *49*;
        dispersion 41–4
posterior 39
pituitary lactotroph *47*, 53, 59–60
cytosolic Ca^{2+} in *58*
poly-L-lysine 144
preovulatory follicle 24–5
pressure dialysis 8
Primaria cultureware 11
progesterone 95
prolactin (PRL) 38, 53, 60
release, regulation by somatostatin
    60
secretion from lactotrophs 53, *58*,
    59–60
prostaglandin E1 76
protein-A coated red blood cells 53
protein kinase C, isozymes of 109

radiocontrast agents 109
rat liver growth factor 120
receptors
damage during tissue digestion 33
FSH 22
hCG 95, 107
IGF type I 107, 120
IGF type 2 111
LH 20, 22, 103, 113
somatostatin 60
TSH 111, 112, 117, 127, 131
reductase
3α 25
5α 25

selenium, supplement to culture medium
    36, 79, 80, 120
Sephadex matrix for pituitary cell
    perifusion 56–7
Sertoli cells 102–4
morphology *102, 104*
somatostatin
release from hypothalamus *136*, 148
supplement to culture medium 120
soybean trypsin inhibitor 97–8, 100
Stadie–Riggs tissue slicer 9
steroidogenic enzymes 20, 22, 80
pseudosubstrate effects 80
steroids, storage of solutions 141
sulphamethazine 109
sulphonylureas 63, 67

testis
contamination of tissue 106
human 96
morphology *98*
porcine 95
rat 95
testosterone
effects on hypothalamic cells 138
secretion from Leydig cells 95, 99,
    105, 105–6
thrombin 75, 80
thrombospondin 112
thyroglobulin 2, 111, 112, 125, 127,
    130
thyroid C-cells 112–13, 128, 130
secretion of calcitonin 112, 128, 130
secretion of somatostatin 112
thyroid endothelial cells 112, *127, 129*
Factor VIII expression 112
thyroid fibroblasts 113, *127, 129*, 130
thyroid follicle 110, 111, *118, 123*
thyroid follicular cells
culture morphology *118*, 123, *124*,
    125
functional assessment 125–7
growth assessment 125
iodide transporter 111, 117
iodide uptake 125–6, *122*
polarisation 111
response to TSH 111, 125–7, 131

thyroid follicular cells (*cont.*)
  heterogeneity in   111
  selection of culture media   118–21
thyroid hormones   111, 112, 125–7
  release   126–7
  synthesis   111
thyroid tissue
  characterisation and morphology
      110–13
  contamination   131
  isolation of cell types
    C-cells   128, 130
    endothelial cells   *127*, 128, *129*
    fibroblasts   *127*, 128, *129*, 130
    follicular cells   117–18, *118*, 121,
        123, *124*, 125–7, 131
  medullary carcinoma   128
  morphology   *115, 116*
  multinodular goitre   114, 125
  processing to obtain cells   114–18
  sources   113–14
    human   113–14
thyroid-stimulating antibodies   109–12
thyroid-stimulating hormone (TSH)   38,
      111, 112, 120, 123–5, 130
  recombinant   120
thyroperoxidase   111, 112, 117, 127
thyrotropin-releasing hormone (TRH)
      50, 57, *136*

tissue disaggregation, consequences of
      incomplete   10
a-tocopherol   75, 80
tolbutamide   67
transferrin, supplement to culture
      medium   75, 80, 99, 100, 120
transforming growth factor-ß (TGF-ß)
      76
trypan blue exclusion test   27, 29, 30,
      32, 34, 43, 79, 101, 144
trypsin   41, 43, 48, 57, 71, 81, 128
  cell damage due to   48, 57
Tygon tubing   55
tyrosine hydroxylase, hypothalamic   138,
      139

Ultroser G   36

vasopressin   75, *137*
verapamil   72
vitamin   C
  as an antioxidant   107
  as supplement to culture medium   99,
      100
vitamin D   6, 7
vitamin E, supplement to culture medium
      36, 99, 100, 107

zymogen granules   67

Printed in the United States
By Bookmasters